Lecture Notes in Computer Science 4935

Commenced Publication in 1973
Founding and Former Series Editors:
Gerhard Goos, Juris Hartmanis, and Jan van Leeuwen

T0223083

Barbara Chapman   Weimin Zheng
Guang R. Gao   Mitsuhisa Sato
Eduard Ayguadé   Dongsheng Wang (Eds.)

A Practical Programming Model for the Multi-Core Era

3rd International Workshop
on OpenMP, IWOMP 2007
Beijing, China, June 3-7, 2007
Proceedings

 Springer

Volume Editors

Barbara Chapman
University of Houston, TX, USA
E-mail: bmchapman@earthlink.net

Weimin Zheng
Tsinghua University, Beijing, China
E-mail: zwm-dcs@tsinghua.edu.cn

Guang R. Gao
University of Delaware, Newark, DE, USA
E-mail: ggao@capsl.udel.edu

Mitsuhisa Sato
University of Tsukuba, Japan
E-mail: msato@ccs.tsukuba.ac.jp

Eduard Ayguadé
Technical University of Catalunya (UPC), Barcelona, Spain
E-mail: eduard@ac.upc.edu

Dongsheng Wang
Tsinghua University, Beijing, China
E-mail: wds@tsinghua.edu.cn

Library of Congress Control Number: 2008928764

CR Subject Classification (1998): D.1.3, D.1, D.2, F.2, G.1-4, J.2, I.6

LNCS Sublibrary: SL 1 – Theoretical Computer Science and General Issues

ISSN   0302-9743
ISBN-10   3-540-69302-5 Springer Berlin Heidelberg New York
ISBN-13   978-3-540-69302-4 Springer Berlin Heidelberg New York

Springer is a part of Springer Science+Business Media

springer.com

© Springer-Verlag Berlin Heidelberg 2008

Typesetting: Camera-ready by author, data conversion by Scientific Publishing Services, Chennai, India
Printed on acid-free paper   SPIN: 12281428   06/3180   5 4 3 2 1 0

Preface

The Third International Workshop on OpenMP, IWOMP 2007, was held at Beijing, China. This year's workshop continued its tradition of being the premier opportunity to learn more about OpenMP, to obtain practical experience and to interact with OpenMP users and developers. The workshop also served as a forum for presenting insights gained by practical experience, as well as research ideas and results related to OpenMP.

A total of 28 submissions were received in response to a call for papers. Each submission was evaluated by three reviewers and additional reviews were received for some papers. Based on the feedback received, 22 papers were accepted for inclusion in the proceedings. Of the 22 papers, 14 were accepted as full papers. We also accepted eight short papers, for each of which there was an opportunity to give a short presentation at the workshop, followed by poster demonstrations. Each paper was judged according to its originality, innovation, readability, and relevance to the expected audience. Due to the limited scope and time of the workshop and the high number of submissions received, only 50% of the total submissions were able to be included in the final program.

In addition to the contributed papers, the IWOMP 2007 program featured several keynote and banquet speakers: Trevor Mudge, Randy Brown, and Shah, Sanjiv. These speakers were selected due to their significant contributions and reputation in the field. A tutorial session and labs were also associated with IWOMP 2007.

We are deeply grateful to the Program Committee members. The large number of submissions received and the diverse topics and coverage of the topics made this review process a particular challenging one. Also, the Program Committee was working under a very tight schedule. We wish to thank all Program Committee members for their assistance in organizing the IWOMP program and determining the paper selection guidelines. Without the solid work and dedication of these committee members, the success of this program, and ultimately the workshop itself, would not have been possible.

We appreciate the contribution of Wenguang Chen and his local team at Tsinghua University, Beijing—in particular, the Local Chair Wenguang Chen—who organized and handled the workshop website for paper submission and review. We also wish to thank the contributions of members of the Organization Committee. We acknowledge the solid work by Dieter an Mey, Ruud van der Pas and Wei Xue for their dedication in organizing the lab, tutorial and workshop sessions. We thank the publicity Co-chairs Federico Massaioli and Dongsheng Wang for their hard work to publicize the IWOMP information under very tight schedule constraints, and to all those who helped with the organization of the final program, the design and maintenance of the workshop websites, the solicitation of sponsorships and support, and numerous other matters related to the local arrangement of the conferences. We also thank the Publication Co-chair Eduard Ayguadé for his work on a special issue based on papers from IWOMP. We are deeply impressed by the efficiency, professionalism and dedication of all of this work.

We would also like to express our gratitude for the support and sponsorship we have received from Intel, Sun Microsystems, the OpenMP Architecture Review Board (ARB), Tsinghua University and University of Delaware. Finally, we give special thanks to Ruini Xue, Liangping Lv, Wenhong Ke at Tsinghua University and Liping Xue and Long Chen at the University of Delaware for their assistance in this endeavor.

April 2008

Barbara Chapman
Weimin Zheng
Bronis R. de Supinski
Guang R. Gao
Mitsuhisa Sato

Organization

Steering Committee

Bronis R. de Supinski	Chair of the Steering Committee, NNSA ASC, LLNL, USA
Dieter an Mey	CCC, RWTH Aachen University, Germany
Eduard Ayguade	Barcelona Supercomputing Center (BSC), Spain
Mark Bull	EPCC, UK
Barbara Chapman	CEO of cOMPunity, USA
Guang R. Gao	University of Delaware, USA
Rudi Eigenmann	Purdue University, USA
Michael Krajecki	University of Reims, France
Ricky A. Kendall	ORNL, USA
Rick Kufrin	NCSA, USA
Federico Massaioli	CASPUR, Italy
Larry Meadows	Intel, USA
Matthias Mueller	ZIH, TU Dresden, Germany
Arnaud Renard	University of Reims, France
Mitsuhisa Sato	University of Tsukuba, Japan
Sanjiv Shah	Intel, OpenMP CEO, USA
Ruud van der Pas	Sun Microsystems, Netherlands
Matthijs van Waveren	Fujitsu, Germany
Michael Wong	IBM, USA
Weimin Zheng	Tsinghua University, China

Organizing Committee

General Co-chairs

Barbara Chapman	University of Houston, USA
Weimin Zheng	Tsinghua University, China

Publicity Co-chairs

Wenguang Chen	Tsinghua University, China
Federico Massaioli	CASPUR, Rome, Italy
Dongsheng Wang	Tsinghua University, China

Publication Co-chairs

Eduard Ayguadé	Barcelona Supercomputing Center (BSC), Spain
Dongsheng Wang	Tsinghua University, China

Short Paper/Poster Session

Chair
Matthias Mueller   ZIH, TU Dresden, Germany

Local Co-chair
Junqing Yu   Huazhong University of Science and Technology,
                                China

Laboratory

Chair
Dieter an May   CCC, RWTH Aachen University, Germany

Local Co-chair
Wei Xue   Tsinghua University, China

Vendor Session

Chair
Ruud van der Pas   Sun Microsystems

Local Co-chair
Dongsheng Wang   Tsinghua University, China

Local Arrangements Chair

Wenguang Chen   Tsinghua University, China

Program Committee

Chair
Guang R. Gao   University of Delaware, USA

Vice Chair
Mitsuhisa Sato   University of Tsukuba, Japan

Members
Dieter an Mey   RWTH Aachen University, Germany
Roch Archambault   IBM, USA
Eduard Ayguadé   Barcelona Supercomputing Center (BSC), Spain
Wenguang Chen   Tsinghua University, China
Nawal Copty   Sun, USA
Luiz DeRose   Cray Inc., USA
Bronis R. de Supinski   LLNL, USA
Rudi Eigenmann   Purdue University, USA
Xiaobing Feng   Institute of Computing Technology, CAS, China
Guang R. Gao   University of Delaware, USA
Hironori Kasahara   University of Waseda, Japan
Ricky A. Kendall   ORNL, USA

Table of Contents

A Proposal for Task Parallelism in OpenMP

Eduard Ayguadé[1], Nawal Copty[2], Alejandro Duran[1], Jay Hoeflinger[3],
Yuan Lin[2], Federico Massaioli[4], Ernesto Su[3], Priya Unnikrishnan[5],
and Guansong Zhang[5]

[1] BSC-UPC
[2] Sun Microsystems
[3] Intel
[4] CASPUR
[5] IBM

Abstract. This paper presents a novel proposal to define task parallelism in OpenMP. Task parallelism has been lacking in the OpenMP
language for a number of years already. As we show, this makes certain kinds of applications difficult to parallelize, inefficient or both. A
subcommittee of the OpenMP language committee, with representatives
from a number of organizations, prepared this proposal to give OpenMP
a way to handle unstructured parallelism. While defining the proposal we
had three design goals: *simplicity of use*, *simplicity of specification* and
consistency with the rest of OpenMP. Unfortunately, these goals were in
conflict many times during our discussions. The paper describes the proposal, some of the problems we faced, the different alternatives, and the
rationale for our choices. We show how to use the proposal to parallelize
some of the classical examples of task parallelism, like pointer chasing
and recursive functions.

1 Introduction

OpenMP grew out of the need to standardize the directive languages of several
vendors in the 1990s. It was structured around parallel loops and meant to handle
dense numerical applications. But, as the sophistication of parallel programmers
has grown and the complexity of their applications has increased, the need for a
less structured way to express parallelism with OpenMP has grown. Users now
need a way to simply identify units of independent work, leaving the decision
about scheduling them to the runtime system. This model is typically called
"tasking" and has been embodied in a number of projects, for example Cilk [1].

The demonstrated feasibility of previous OpenMP-based tasking extensions
(for example workqueueing [2] and dynamic sections [3]) combined with the
desire of users to standardize it, has caused one of the priorities of the OpenMP
3.0 effort to be defining a standardized tasking dialect. In September of 2005, the
tasking subcommittee of the OpenMP 3.0 language committee began meeting,
with a goal of defining this tasking dialect. Representatives from Intel, UPC,
IBM, Sun, CASPUR and PGI formed the core of the subcommittee.

B. Chapman et al. (Eds.): IWOMP 2007, LNCS 4935, pp. 1–12, 2008.

This paper, written by some of the tasking subcommittee members, is a description of the resulting tasking proposal, including our motivations, our goals for the effort, the design principles we attempted to follow, the tough decisions we had to make and our reasons for them. We will also present examples to illustrate how tasks are written and used under this proposal.

2   Motivation and Related Work

OpenMP "is somewhat tailored for large array-based applications"[4]. This is evident in the limitations of the two principal mechanisms to distribute work among threads. In the *loop* construct, the number of iterations must be determined on entry to the loop and cannot be changed during its execution. In the **sections** construct, the sections are statically defined at compile time.

A common operation like a dynamic linked list traversal is thus difficult to parallelize in OpenMP. A possible approach, namely the transformation at run time of the list to an array, as shown in fig. 1, pays the overheads of the array construction, which is not easy to parallelize. Another approach, using the **single nowait** construct as shown in fig. 2, can be used. While elegant, it's non-intuitive, and inefficient because of the usually high cost of the **single** construct[5].

```
1   p = listhead;
2   num_elements=0;
3   while (p) {
4       list_item[num_elements++]=p;
5       p=next(p);
6   }
7   #pragma omp parallel for
8   for (int i=0;i<num_elements;i++)
9       process(list_item[i]);
```

```
#pragma omp parallel private (p)
{
    p = listhead;
    while(p) {
        #pragma omp single nowait
            process(p);
        p = next(p);
    }
}
```

Fig. 1. Parallel pointer chasing with the *inspector-executor* model

Fig. 2. Parallel pointer chasing using **single nowait**

Both techniques above lack generality and flexibility. Many applications (ranging from document bases indexing to adaptive mesh refinement) have a lot of potential concurrency, which is not regular in form, and varies with processed data. Dynamic generation of different units of work, to be asynchronously executed, allows one to express irregular parallelism, to the benefit of performance and program structure. Nested parallelism could be used to this aim, but at the price of significant performance impacts and increased synchronizations.

The present OpenMP standard also lacks the ability to specify structured dependencies among different units of work. The **ordered** construct assumes a sequential ordering of the activities. The other OpenMP synchronization constructs, like **barrier**, actually synchronize the whole team of threads, not work units. This is a significant limitation on the coding of hierarchical algorithms like those used in tree data structure traversal, multiblock grid solvers, adaptive mesh refinement[6], dense linear algebra [7,8,9] (to name a few). In principle, nested parallelism could be used to address this issue, as in the example shown in fig. 3. However, overheads in parallel region creation, risks of oversubscribing system resources, difficulties in load balancing, different behaviors of different implementations, make this approach impractical.

```
1    void traverse(binarytree *p) {
2      #pragma omp parallel sections num_threads(2)
3      {
4        #pragma omp section
5          if (p->left) traverse(p->left);
6        #pragma omp section
7          if (p->right) traverse(p->right);
8      }
9      process(p);
10   }
```

Fig. 3. Parallel depth-first tree traversal

The Cilk programming language[1] is an elegant, simple, and effective extension of C for multithreading, based on dynamic generation of tasks. It is important and instructive, particularly because of the *work-first* principle and the *work-stealing* technique adopted. However, it lacks most of the features that make OpenMP very efficient in many computational problems.

The need to support irregular forms of parallelism in HPC is evident in features being included in new programming languages, notably X10 [10] (*activities* and *futures*), and Chapel [11] (the cobegin statement).

The Intel *workqueueing* model [2] was the first attempt to add dynamic task generation to OpenMP. This proprietary extension allows hierarchical generation of tasks by nesting of taskq constructs. Synchronization of descendant tasks is controlled by means of implicit barriers at the end of taskq constructs. The implementation exhibits some performance issues [5,8].

The Nanos group at UPC proposed *Dynamic Sections* as an extension to the standard sections construct to allow dynamic generation of tasks [3]. Direct nesting of section blocks is allowed, but hierarchical synchronization of tasks can only be attained by nesting of a parallel region. The Nanos group also proposed the pred and succ constructs to specify precedence relations among statically named sections in OpenMP [12]. These are issues that may be explored as part of our future work.

The point-to-point synchronization explored at EPCC [13] improves in performance and flexibility with respect to OpenMP barrier. However, the synchronization still involves threads, not units of work.

3   Task Proposal

The current OpenMP specification (version 2.5) is based on threads. The execution model is based on the fork-join model of parallel execution where all threads have access to a shared memory. Usually, threads share work units through the use of worksharing constructs. Each work unit is bound to a specific thread for its whole lifetime. The work units use the data environment of the thread they are bound to.

Our proposal allows the programmer to specify deferrable units of work that we call *tasks*. Tasks, unlike current work units, are not bound to a specific thread. There is no a priori knowledge of which thread will execute a task. Task execution may be deferred to a later time, and different parts of a task may be executed by different threads. As well, tasks do not use the data environment of any thread but have a data environment of their own.

3.1   Terminology

Task. A structured block, executed once for each time the associated task construct is encountered by a thread. The task has private memory associated with it that stays persistent during a single execution. The code in one instance of a task is executed sequentially.

Suspend/resume point. A suspend point is a point in the execution of the task where a thread may suspend its execution and run another task. Its corresponding resume point is the point in the execution of the task that immediately follows the suspend point in the logical code sequence. Thus, the execution that is interrupted at a suspend point resumes at the matching resume point.

Thread switching. A property of the task that allows its execution to be suspended by one thread and resumed by a different thread, across a suspend/resume point. Thread switching is disabled by default.

3.2   The Task Construct

The C/C++ syntax[1] is as follows:

```
#pragma omp task [clause[[,]clause] ...]
        structured-block
```

The thread that encounters the task construct creates a task for the associated structured block, but its execution may be deferred until later. The execution gives rise to a task region and may be done by any thread in the current parallel team.

A task region may be nested inside another task region, but the inner one is not considered part of the outer one. The task construct can also be specified inside any other OpenMP construct or outside any explicit parallel construct. A task is guaranteed to be complete before the next associated thread or task barrier completes.

---

[1] Fortran syntax is not shown in this paper because of space limitations.

The optional clauses can be chose from:

- `untied`
- `shared (variable-list)`
- `captureprivate (variable-list)`
- `private (variable-list)`

The `untied` clause enables thread switching for this task. A task does not inherit the effect of a `untied` clause from any outer task construct.

The remaining three clauses are related to data allocation and initialization for a task. Variables listed in the clauses must exist in the enclosing scope, where each is referred to as the *original* variable for the variable in the task.

References within a task to a variable listed in its `shared` clause refer to the original variable. New storage is created for each `captureprivate` variable and initialized to the original variable's value. All references to the original variable in the task are replaced by references to the new storage. Similarly, new storage is created for `private` variables and all references to them in the task refer to the new storage. However, `private` variables are not automatically initialized.

The default for all variables with implicitly-determined sharing attributes is `captureprivate`.

3.3   Synchronization Constructs

Two constructs (`taskgroup` and `taskwait`) are provided for synchronizing the execution of tasks.

The C/C++ syntax for `taskgroup` and `taskwait` is as follows:

```
#pragma omp taskgroup    │    #pragma omp taskwait
        structured-block │
```

The `taskgroup` construct specifies that execution will not proceed beyond its associated structured block until all direct descendant tasks generated within it are complete. The `taskwait` construct specifies that execution will not proceed beyond it until all direct descendant tasks generated within the current task, up to that point, are complete.

There is no restriction on where a `taskgroup` construct or a `taskwait` construct can be placed in an OpenMP program. `Taskgroup` constructs can be nested.

3.4   Other Constructs

When a thread encounters the `taskyield` construct, it is allowed to look for another available task to execute. This directive becomes an explicit suspend/resume point in the code. The syntax for C/C++ is:

```
#pragma omp taskyield
```

3.5 OpenMP Modifications

The tasking model requires various modifications in the OpenMP 2.5 specification. We list some of these below.

– **Execution Model.** In the 2.5 specification, there is no concept of the deferral of work or of suspend or resume points. In our proposal, a task may be suspended/resumed and execution of the task may be deferred until a later time. Moreover, a task may be executed by different threads during its lifetime, if thread switching is explicitly enabled.
– **Memory Model.** In the 2.5 specification, a variable can be shared among threads or be private to a thread. In the tasking model, a variable can be shared among tasks, or private to a task. The default data sharing attributes for tasks differs from the defaults for `parallel` constructs.
– **Threadprivate data and properties.** Inside OpenMP 2.5 work units, `threadprivate` variables and thread properties (e.g. thread id) can be used safely. Within tasks, this may not be the case. If thread switching is enabled, the executing thread may change across suspend or resume points, so thread id and `threadprivate` storage may change.
– **Thread Barrier.** In our proposal, a task is guaranteed to be complete before the associated thread or task barrier completes. This gives a thread barrier extra semantics that did not exist in the 2.5 specification.
– **Locks.** In the 2.5 specification, locks are owned by threads. In our proposal, locks are owned by tasks.

4 Design Principles

Unlike the structured parallelism currently exploited using OpenMP, the tasking model is capable of exploiting irregular parallelism in the presence of complicated control structures. One of our primary goals was to design a model that is easy for a novice OpenMP user to use and one that provides a smooth transition for seasoned OpenMP programmers. We strived for the following as our main design principles: *simplicity of use, simplicity of specification* and *consistency with the rest of OpenMP*, all without losing the expressiveness of the model. In this section, we outline some of the major decisions we faced and the rationale for our choices, based on available options, their trade-offs and our design goals.

What Form Should the Tasking Construct(s) Take?

Option 1. *A new work-sharing construct pair*: It seemed like a natural extension of OpenMP to use a work-sharing construct analogous to `sections` to set up the data environment for tasking and a `task` construct analogous to `section` to define a task. Under this scheme, tasks would be bound to the work-sharing construct. However, these constructs would inherit all the restrictions applicable to work-sharing constructs, such as a restriction against nesting them. Because of the dynamic nature of tasks, we felt that this would place unnecessary restrictions on the applicability of tasks and interfere with the basic goal of using tasks for irregular computations.

Option 2. *One new OpenMP construct*: The other option was to define a single task construct which could be placed anywhere in the program and which would cause a task to be generated each time a thread encounters it. Tasks would not be bound to any specific OpenMP constructs. This makes tasking a very powerful tool and opens up new parallel application areas, previously unavailable to the user due to language limitations. Also, using a single tasking construct significantly reduces the complexity of construct nesting rules. The flexibility of this option seemed to make it the most easy to merge into the rest of OpenMP, so this was our choice.

Should We Allow Thread Switching and What Form Should It Take?

Thread switching is a concept that is alien to traditional OpenMP. In OpenMP, a thread always executes a unit of work in a work-share from start to finish. This has encouraged people to use the thread number to identify a unit of work, and to store temporary data in **threadprivate** storage. This has worked very well for the kind of parallelism that can be exploited in regular control structures. However, tasking is made for irregular parallelism. Tasks can take wildly differing amounts of execution time. It is very possible that the only thread eligible to generate tasks might get stuck executing a very long task. If this happens, the other threads in the team could be sitting idle, waiting for more tasks, and starvation results. Thread switching would enable one of the waiting threads to take over task-generation in a situation like this.

Thread switching provides greater flexibility and potentially higher performance. It allows *work-first* execution of tasks, where a thread immediately executes the encountered task and another thread continues where the first thread left off. It has been shown that work-first execution can result in better cache reuse [1]. However, switching to a new thread changes the thread number and any *threadprivate* data being used, which could be surprising to an experienced OpenMP programmer.

Balancing the benefits of thread switching against the drawbacks, it was decided to allow thread switching in a limited and controlled manner. We decided to disable thread switching by default. It is only allowed inside tasks using the **untied** clause. Without the **untied** clause, the programmer can depend on the thread number and **threadprivate** data in a task just as in other parts of OpenMP.

Should the Implementation Guarantee That Task References to Stack Data Are Safe?

A **task** is likely to have references to the data on the stack of the routine where the **task** construct appears. Since the execution of a task is not required to be finished until the next associated task barrier, it is possible that a given task will not execute until after the stack of the routine where it appears is already popped and the stack data over-written, destroying local data listed as **shared** by the task.

The committee's original decision was to require the implementation to guarantee stack safety by inserting task barriers where required. We soon realized that there are circumstances where it is impossible to determine at compile time exactly when execution will leave a given routine. This could be due to a complex branching structure in the code, but worse would be the use of `setjmp/longjmp`, C++ exceptions, or even vendor-specific routines that unwind the stack. When you add to this the problem of the compiler understanding when a given pointer dereference is referring to the stack (even through a pointer argument to the routine), you find that in a significant number of cases the implementation would be forced to conservatively insert a task barrier immediately after many task constructs, severely restricting the parallelism possible with tasks.

Our decision was to simply state that it is the user's responsibility to insert any needed task barriers to provide any needed stack safety.

What Should Be the Default Data-Sharing Attributes for Variables in Tasks? Data-sharing attributes for variables can be pre-determined, implicitly determined or explicitly determined. Variables in a task that have pre-determined sharing attributes are not allowed in clauses, and explicitly-determined variables do not need defaults, by definition. However, the data-sharing attributes for implicitly-determined variables require defaults.

The sharing attributes of a variable are strongly linked to the way in which it is used. If a variable is shared among a thread team and a task must modify its value, then the variable should be **shared** on the task construct and care must be taken to make sure that fetches of the variable outside the task wait for the value to be written. If the variable is read-only in the task, then the safest thing would be to make the variable **captureprivate**, to make sure that it will not be deallocated before it is used. Since we decided to not guarantee stack safety for tasks, we faced a hard choice:

Option 1 make data primarily **shared**, analogous to using **shared** in the rest of OpenMP. This choice is consistent with existing OpenMP. But, with this default, the danger of data going out of scope is very high. This would put a heavy burden on the user to ensure that all the data remains allocated while it is used in the task. Debugging can be a nightmare for things that are sometimes deallocated prematurely.

Option 2 make data primarily **captureprivate**. The biggest advantage of this choice is that it minimizes the "data-deallocation"' problem. The user only needs to worry about maintaining allocation of variables that are explicitly **shared**. The downside to using **captureprivate** as the default is that Fortran parameters and C++ reference parameters will, by default, be captured by tasks. This could lead to errors when a task writes into reference parameters.

Option 3 make some variables **shared** and some **captureprivate**. With this choice, the rules could become very complicated, and with complicated rules the chances for error increase. The most likely effect would be to force the programmer to explicitly place all variables in some clause, as if there were no defaults at all.

In the end, we decided to make all variables with implicitly-determined sharing attributes default to `captureprivate`. While not perfect, this choice gives programmers the most safety, while not being overly complex.

5 Examples of Use

In this section, we revisit examples we used in section 2, and write those applications based on our task proposal.

5.1 Pointer Chasing in Parallel

Pointer chasing (or *pointer following*) can be described as a segment of code working on a list of data items linked by pointers. When there is no dependence, the process can be executed in parallel.

We already described in Section 2 a couple of non-efficient or non-intuitive parallelization strategies for this kernel (fig. 1 and 2). In addition, both solutions fail if the list itself needs to be updated dynamically during the execution. In fact, with proper synchronization, the second solution may append more items at the end of the link, while the first one will not work at all.

All these problems go away with the new task proposal. This simple kernel could be parallelized as shown in fig. 4.

```
1   #pragma omp parallel
2   {
3       #pragma omp single
4       {
5           p = listhead;
6           while(p) {
7               #pragma omp task
8                   process(p)
9               p=next(p);
10          }
11      }
12  }
```

Fig. 4. Parallel pointer chasing using `task`

```
#pragma omp parallel private (p)
{
    #pragma omp for
    for (int i=0; i< num_lists; i++) {
        p = listheads[i];
        while(p) {
            #pragma omp task
                process(p)
            p=next(p);
        }
    }
}
```

Fig. 5. Parallel pointer chasing on multiple lists using `task`

The `task` construct gives more freedom for scheduling (see the next section for more). It is also more straightforward to add a synchronization mechanism to work on a dynamically updated list[2].

In fig. 4, the `single` construct ensures that only one thread will encounter the `task` directive and start a task nest. It is also possible to use other OpenMP constructs, such as `sections` and `master`, or even an `if` statement using thread id, to achieve the same effect.

---

[2] We will leave it as an exercise for readers to write the `next` function in the different versions.

```
 1 int fib(int n) {
 2    int x, y, sum;
 3    if (n<2)
 4      return n;
 5 #pragma omp taskgroup
 6    {
 7 #pragma omp task shared(x)
 8       x=fib(n-1);
 9 #pragma omp task shared(y)
10       y=fib(n-2);
11    }
12    return x+y;
13 }
```

Fig. 6. Fibonacci with recursive `task`

More interestingly, we may have multiple lists to be processed simultaneously by the threads of the team, as in fig. 5. This results in better load balancing when the number of lists does not match the number of threads, or when the lists have very different lengths.

5.2   Recursive Task

Another scenario of using the `task` directive is shown in fig. 6, inspired by one of the examples in the Cilk project [1]. This code segment calculates the Fibonacci number. If a call to this function is encountered by a single thread in a parallel region, a nested task region will be spawned to carry out the computation in parallel. The TASKGROUP construct defines a code region which the encountering thread should wait for. Both i and j are on the stack of the parent function that invokes the new tasks. Please refer to section 3.3 for the details. Notice that although OpenMP constructs are combined with a recursive function, it is still equivalent to its sequential counterpart.

Other applications falling into this category include: virus shell assembly, graphics rendering, n-body simulation, heuristic search, dense and sparse matrix computation, friction-stir welding simulation and artificial evolution.

We can also rewrite the example in fig. 3 as in fig. 7. In this figure, we use `task` to avoid the nested `parallel` regions. Also, we can use a flag to make the post order processing optional.

6   Future Work

So far, we have presented a proposal to seamlessly integrate task parallelism into the current OpenMP standard. The proposal covers the basic aspects of task parallelism, but other areas are not covered by the current proposal and may be subject of future work. One such possible extension is a reduction operation performed by multiple tasks. Another is specification of dependencies between tasks, or point-to-point synchronizations among tasks. These extensions may

```
1 void traverse(binarytree *p, bool postorder) {
2   #pragma omp task
3     if (p->left) traverse(p->left, postorder);
4   #pragma omp task
5     if (p->right) traverse(p->right, postorder);
6   if (postorder) {
7     #pragma omp taskwait
8   }
9   process(p);
10 }
```

Fig. 7. Parallel depth-first tree traversal

be particularly important for dealing with applications that can be expressed through a task graph or that use pipelines.

The task proposal allows a lot of freedom for the runtime library to schedule tasks. Several simple strategies for scheduling tasks exist but it is not clear which will be better for the different target applications as these strategies have been developed in the context of recursive applications. Furthermore, more complex scheduling strategies can be developed that take into account characteristics of the application which can be found either at compile time or run time. Another option would be developing language changes that allow the programmer to have greater control of the scheduling of tasks so they can implement complex schedules (e.g. shortest job time, round robin) [8].

7   Conclusions

We have presented the work of the tasking subcommittee: a proposal to integrate task parallelism into the OpenMP specification. This allows programmers to parallelize program structures like `while` loops and recursive functions more easily and efficiently. We have shown that, in fact, these structures are easy to parallelize with the new proposal.

The process of defining the proposal has not been without difficult decisions, as we tried to achieve conflicting goals: *simplicity of use, simplicity of specification* and *consistency with the rest of OpenMP*. Our discussions identified trade-offs between the goals, and our decisions reflected our best judgments of the relative merits of each. We also described how some parts of the current specification need to be changed to accommodate our proposal.

In the end, however, we feel that we have devised a balanced, flexible, and very expressive dialect for expressing unstructured parallelism in OpenMP programs.

Acknowledgments

The authors would like to acknowledge the rest of participants in the tasking subcommittee (Brian Bliss, Mark Bull, Eric Duncan, Roger Ferrer, Grant Haab,

Diana King, Kelvin Li, Xavier Martorell, Tim Mattson, Jeff Olivier, Paul Petersen, Sanjiv Shah, Raul Silvera, Xavier Teruel, Matthijs van Waveren and Michael Wolfe) and the language committee members for their contributions to this tasking proposal. The Nanos group at BSC-UPC is supported has been supported by the Ministry of Education of Spain under contract TIN2007-60625, and the European Commission in the context of the SARC integrated project #27648 (FP6).

References

1. Frigo, M., Leiserson, C.E., Randall, K.H.: The implementation of the Cilk-5 multi-threaded language. In: PLDI 1998. Proceedings of the ACM SIGPLAN 1998 conference on Programming language design and implementation, pp. 212–223. ACM Press, New York (1998)
2. Shah, S., Haab, G., Petersen, P., Throop, J.: Flexible control structures for parallellism in OpenMP. In: 1st European Workshop on OpenMP (September 1999)
3. Balart, J., Duran, A., Gonzàlez, M., Martorell, X., Ayguadé, E., Labarta, J.: Nanos Mercurium: A Research Compiler for OpenMP. In: 6th European Workshop on OpenMP (EWOMP 2004), September 2004, pp. 103–109 (2004)
4. OpenMP Architecture Review Board. OpenMP Application Program Interface (May 2005)
5. Massaioli, F., Castiglione, F., Bernaschi, M.: OpenMP parallelization of agent-based models. Parallel Computing 31(10-12), 1066–1081 (2005)
6. Blikberg, R., Sørevik, T.: Load balancing and OpenMP implementation of nested parallelism. Parallel Computing 31(10-12), 984–998 (2005)
7. Salvini, S.: Unlocking the Power of OpenMP. In: 5th European Workshop on OpenMP (EWOMP 2003) (September 2003)
8. Van Zee, F.G., Bientinesi, P., Low, T.M., van de Geijn, R.A.: Scalable Parallelization of FLAME Code via the Workqueuing Model. ACM Trans. Math. Soft. (submitted, 2006)
9. Kurzak, J., Dongarra, J.: Implementing Linear Algebra Routines on Multi-Core Processors with Pipelining and a Look Ahead. LAPACK Working Note 178, Dept. of Computer Science, University of Tennessee (September 2006)
10. The X10 Design Team. Report on the Experimental Language X10. Technical report, IBM (February 2006)
11. Callahan, D., Chamberlain, B.L., Zima, H.P.: The Cascade High Productivity Language. In: 9th Int. Workshop on High-Level Parallel Programming Models and Supportive Environments (HIPS 2004), pp. 52–60. IEEE Computer Society, Los Alamitos (2004)
12. Gonzalez, M., Ayguadé, E., Martorell, X., Labarta, J.: Exploiting pipelined executions in OpenMP. In: 32nd Annual International Conference on Parallel Processing (ICPP 2003) (October 2003)
13. Bull, J.M., Ball, C.: Point-to-Point Synchronisation on Shared Memory Architectures. In: 5th European Workshop on OpenMP (EWOMP 2003) (September 2003)

Support for Fine Grained Dependent Tasks in OpenMP

Oliver Sinnen, Jsun Pe, and Alexei Vladimirovich Kozlov

Department of Electrical and Computer Engineering
University of Auckland
Private Bag 92019
Auckland, New Zealand
o.sinnen@auckland.ac.nz

Abstract. OpenMP is widely used for shared memory parallel programming and is especially useful for the parallelisation of loops. When it comes to task parallelism, however, OpenMP is less powerful and the sections construct lacks support for dependences and fine grained tasks. This paper proposes a new work-sharing construct, tasks, which is a generalisation of sections. It goes beyond sections by allowing unbalanced and finer grained tasks with arbitrary dependence structure. A proof-of-concept compiler has been implemented for the new directives, which includes a state-of-the-art scheduling algorithm for task graphs. Experiments with a large set of programs were conducted using the new directives. The results demonstrate that the new approach can efficiently exploit the task parallelism inherent in the code, without introducing any additional overhead.

1 Introduction

For shared memory parallel programming with the languages Fortran and C/C++, OpenMP has long been a de facto standard [1]. Augmenting the Fortran or C/C++ source code with OpenMP's work-sharing constructs is very powerful for the parallelisation of iterative computations, i.e. loops. For task parallelism, OpenMP offers the sections construct, which allows to define independent sections that can be executed concurrently. The underlying notion is that each section will be executed by a different thread.

This implies that each section of the sections construct is rather large. Further, to achieve a good load balance, and hence to use the processors efficiently, the computational load of each section must be similar. But most importantly, there can be no dependences between the sections, or the programmer must synchronise the corresponding parts manually with the usual mechanisms.

This paper proposes an extension to OpenMP and presents a proof-of-concept implementation that allows to easily parallelise fine grained tasks with arbitrary dependence structure. Analogue to the sections-section construct, the new directives are designated tasks-task and can be considered a generalisation of the sections-section concept. The task blocks of a tasks construct need not to have balanced load. Our proof-of-concept source-to-source compiler schedules the task blocks automatically with a state-of-the-art task scheduling algorithms. Like all

B. Chapman et al. (Eds.): IWOMP 2007, LNCS 4935, pp. 13–24, 2008.

OpenMP directives, the new `tasks-task` directives maintain the sequential semantic of the program. The approach addresses a different type of parallelism than the `taskq-task` work-queueing concept [8], as implemented in the Intel OpenMP compiler [11]. Having said this, it is conceivable that the two concepts can later be unified.

Experimental results demonstrate that this approach can exploit the inherent parallelism of the code. It also shows that there is no additional overhead in comparisons to the `sections-section` directives.

The rest of this paper is organised as follows. Section 2 introduces the syntax and semantics of the new `tasks-task` directives. The structure and implementation of the proof-of-concept compiler are discussed in Section 3. An experimental evaluation of the new approach is given in Section 4 and the paper concludes in Section 5.

2   New `tasks-task` Directives

This section proposes a new work-sharing construct `tasks` for OpenMP. In essence, it is a generalisation of the existing `sections` construct [1], but is more powerful in three aspects: i) it allows the parallelisation of partially dependent tasks; ii) it is more scalable; and iii) it enables more efficient scheduling. The `tasks` directive identifies a non-iterative work-sharing construct. It has the following syntax:

```
#pragma omp tasks [clause[[,]clause...]
structured-block
```

where `clause` can be any of the following, which are almost identical to the clauses permitted by the `sections` construct:

```
private(variable-list)
firstprivate(variable-list)
lastprivate(variable-list)
reduction(operator; variable-list)
```

The combined `parallel tasks` construct is defined as for the `sections` construct. A `tasks` directive encloses one or more `task` directives. Each `task` directive encloses a structured block of code that has to be executed in sequential, i.e. by a single thread. Different `task` blocks, however, can be executed in parallel, if their dependences permit. More on the dependences later, let us first look at the syntax of the `task`:

```
#pragma omp task name weight [dependsOn(list-of-tasks)]
structured-block
```

In contrast to a `section`, there are three additional elements:

1. name: An identifier of this `task`, e.g. A or `job1`.
2. weight: An integer value that describes the relative computational weight of the `task`. As an example consider two tasks A and B with a `weight` value of 1 and

5, respectively. This means that B's estimated execution time is five times longer than that of A.

3. **dependsOn**: The *list-of-tasks*, which are argument to the dependsOn key word, enumerate the task blocks on which this task depends on. For example, dependsOn(A,C) means that this task depends on the tasks A and C. If the dependsOn keyword is omitted or the *list-of-tasks* is empty, the task is independent of any other task within the enclosing tasks construct.

The task directive must only appear within the lexical extent of the tasks directive. Through the utilisation of the dependsOn statement, a partial order between the task blocks of a tasks construct can be established. The execution order that is established by the OpenMP complier must adhere to precedence constraints established by the dependsOn statements. For the tasks construct, the dependsOn statements substitute and generalise the ordered clause of OpenMP that can be employed with other work-sharing constructs.

Note that the *list-of-tasks* can only contain task identifiers of lexically preceeding task blocks – forward references are not allowed. This makes dependence cycles impossible as it enforces a precedence order of the task blocks. More importantly, it also enforces the usual sequential semantics of OpenMP constructs. Hence, the augmented code retains its sequential semantic. Figure 1 depicts on the left a simple tasks construct. The code blocks of the task directives are symbolised by Block_Code_x comments.

To see the advantage and the power of the tasks-task directives, let us first look back at the semantic of the sections-section directives. In a sections construct, each section is usually associated with at thread. During execution, that commonly means that each section is executed by a different thread. Hence, the programmer needs to manually balance the computational load of the section blocks. Furthermore, code parts that require a certain execution order must be put into the same section or manually synchronised.

In contrast, the semantic of the tasks-task construct has no such association with threads. The programmer should try to define many task blocks – many more than the maximum number of threads that will be available during execution. During compilation, or even at runtime, the task blocks are grouped and ordered for execution. This scheduling takes into account the computational load of the task blocks (as specified with the weight argument), their precedence constraints and the number of threads. Hence, the schedule can be optimised for the number of available processors.

The new constructs are very appropriate for the parallelisation of existing code that features some task parallelism. The programmer defines task blocks and estimates the execution time of each block in the weight value. When a certain block or function must be completed before another one, the programmer simply needs to indicate this in a dependsOn statement. There is no need for reorganising the code structure due to dependences. Note, that the programmer has to analyse the dependences between code parts for any type of parallelisation, there is no additional burden inflicted by the new directives.

3 Compiler Implementation

A proof-of-concept OpenMP compiler was implemented for the new `tasks-task`
directives. In order to avoid the construction of a fully featured compiler in this project,
a source-to-source approach was chosen. The developed compiler's input is a source file
with extended OpenMP directives. During its execution, the compiler converts the new
`tasks-task` directives into standard OpenMP work-sharing constructs. The output
source file can then be compiled with a conventional OpenMP compiler.

The complier architecture can be divided into three parts:

1. Parser (front-end) – parses the source code and constructs a task graph
2. Scheduler (middle-end) – schedules the task graph to a number of processors
3. Code generator (back-end) – creates code from the schedule and the original code

It was implemented for the Java programming language. While there is no official
Java language binding for the OpenMP specification, a source-to-source Java complier
for OpenMP-like directives exists, called JOMP [2,5]. The employed syntax for the
OpenMP directives is almost identical to that of the C/C++ language binding, with
the notable difference that directives are comments in the Java syntax and not pragmas. Hence, a directive is initiated with \\omp instead of the #pragma omp used in
C/C++.

3.1 Parser

For the chosen approach of a source-to-source compiler, the parser can be quite simple.
The main job of the parser front-end is to parse the new directives and to construct a
task graph for each `tasks` work-sharing construct.

Graphs are widely used in parallel computing as an abstraction of a program [9]. For
scheduling, a program $\mathcal{P}$ (or parts of it) is represented by the directed acyclic graph
$G = (\mathbf{V}, \mathbf{E}, w)$, called task graph, where $\mathbf{V}$ is the set of nodes representing the tasks
of $\mathcal{P}$ and $\mathbf{E}$ is the set of edges representing the dependences between the tasks. An
edge $e_{ij} \in \mathbf{E}$ represents the communication from node n_i to node n_j, $n_i, n_j \in \mathbf{V}$. The
positive weight $w(n)$ of node $n \in \mathbf{V}$ represents its computation cost.

In order to generate such a task graph, the parser needs to localise the `task` blocks in
each `tasks` construct. A task graph is created for each `tasks` construct, consisting of
one node n for each of its `task` blocks. The code enclosed in the `task` block is associated with the node n. Each n is given the parsed identifier of the `task` block and $w(n)$
is sent to the corresponding parsed weight. Edges are created from the the `dependsOn`
statements. One edge e is created for each task identifier x in a `dependsOn` statement:
e is directed from the node of x to the node of the `dependsOn` statement.

Figure 1 depicts this parsing procedure for a simple `tasks` construct. The code
blocks of the `task` directives are symbolised by `Block_Code_x` comments. As can
be seen, nodes A-G are created for the seven `task` blocks. The nodes are identified with
the `name` element of the corresponding blocks and have their corresponding `weight`.
For example, the first `task` block corresponds to node A with weight $w(A) = 2$.
As described above, the edges are inserted into the graph based on the `dependsOn`
statements. For example, there are three `dependsOn` statements containing A, thus
there are three edges e_{AB}, e_{AC}, e_{AD} directed from A to the corresponding target nodes.

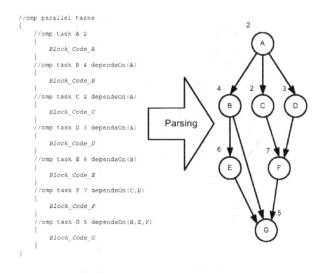

Fig. 1. Constructing a task graph from `tasks-task` directives

3.2   Scheduler

The middle-end of the compiler deals with the scheduling of the `task` blocks. Strictly speaking, this scheduling can be performed later during program execution, implemented in a runtime library. For our proof-of-concept complier, it is performed during compilation for simplicity. An advantage of this approach is that the scheduling costs are paid only once during compilation time.

With the construction of the task graph, the code part to be parallelised (i.e. the `tasks` construct) has been divided into subtasks and their dependences have been established. An additional information necessary to perform the scheduling is the number of processors (or threads). In this implementation, the number is passed to the compiler as an argument. For a runtime implementation, the scheduling can be performed for the automatically detected number of available threads.

Simple dynamic schemes, like a central work-queue, cannot be employed for scheduling due to the dependences between the tasks. Scheduling of task graphs is a difficult problem – in fact it is NP-hard [3] – and many heuristic algorithms have been proposed [9]. An essential aspect of scheduling is the underlying model of the target parallel system. In this proof-of-concept compiler, a classical task scheduling model that does not consider communication costs is employed [9]. The implications of this are discussed later. The following assumptions are made about the target system: i) the system is dedicated, i.e. no other program is executed at the same time; ii) tasks are non-preemptive; iii) communication is cost free. Figure 2 depicts a schedule for the task graph of Figure 1 on two processors under this model. For example, node B can only start on P_2 after A has finished on P_1 due to the dependence expressed through the edge e_{AB}. Note that B starts immediately after A, since the utilised model does not assume any communication costs.

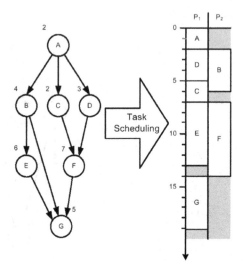

Fig. 2. Scheduling of the task graph

Using this model, a simple, low complexity and generally good scheduling heuristic [4], called list scheduling was selected and implemented. In its simplest form, the first part of list scheduling sorts the nodes $\mathbf{V}$ of the task graph $G = (\mathbf{V}, \mathbf{E}, w)$ according to a priority scheme, while respecting the precedence constraints of the nodes. In the second part, each node of the list is successively scheduled to a processor P chosen for the node. Usually, the chosen processor P is the one of all processors $\mathbf{P}$ that allows the earliest start time of the node. Algorithm 1 outlines this list scheduling procedure.

Algorithm 1. Simple list scheduling
▷ *1. Part:*
Sort nodes $n \in \mathbf{V}$ into list L, according to priority scheme and precedence constraints.
▷ *2. Part:*
for each $n \in L$ **do**
  Choose processor $P \in \mathbf{P}$ for n that allows earliest start time.
  Schedule n on P.
end for

As was shown in many experiments, e.g. [7], the lengths of the schedules produced by list scheduling are strongly influenced by the order in which the nodes are processed, hence by the priority scheme. Sorting the nodes in decreasing (computation) bottom level order is simple and gives usually good results [9]. The computation bottom level $bl_w(n)$ of a node n is the length of the longest path starting with n, where the length of a path is here defined as the sum of all of its node weights. Recursively, this is

$$bl_w(n_i) = w(n_i) + \max_{n_j \in \mathbf{succ}(n_i)} \{bl_w(n_j)\}, \qquad (1)$$

where $\mathbf{succ}(n_i)$ is the set of successor (child) nodes of n_j.

List scheduling using bottom level node order is usually fast as it is straight forward to implement and has near linear complexity $O(|\mathbf{V}|(|\mathbf{P}| + \log|\mathbf{V}|) + |\mathbf{P}||\mathbf{E}|)$ [9]. In any case, the number of nodes (i.e. task blocks) should rarely be high in a tasks construct, hence the complexity should never be important. For that reason, more sophisticated algorithms might be used in the future, in particular algorithms that obtain an optimal solution [10].

3.3   Code Generation

In the last step of the compilation, the code is generated according to the schedule. As this is a source-to-source compiler, the generated code is composed of the original code in the task blocks and standard OpenMP directives.

The simple solution is to use the sections construct of standard OpenMP. Each tasks construct translates to one sections construct. Within the sections construct, one section block is created for each processor P of the schedule. This follows the semantic of sections in that each section will be executed by a different thread. The body of the section block in turn contains the code of the task blocks corresponding to the nodes scheduled on the processor P. This code is arranged in the start time order of the nodes on P in the schedule.

Following this procedure, the code for the example of Figures 1 and 2 looks like this:

```
//omp parallel sections
{
   //omp section
   {
      Block_Code_A
      Block_Code_D
      Block_Code_C
      Block_Code_E
      Block_Code_G
   }
   //omp section
   {
      Block_Code_B
      Block_Code_F
   }
}
```

Synchronisation. While this code accurately represents the mapping and the order of the nodes in the schedule, it does not completely adhere to the precedence constraints established by the edges of the task graph. For example, code block B should not start execution before A has finished (due to edge e_{AB}). However, since they are executed in different section blocks, i.e. threads, there is no guarantee for the correct execution order. Note that the precedence order between code blocks in the same section is imposed through the lexical order of the code.

To adhere to the precedence constraints of code blocks in different section constructs, a synchronisation mechanism is necessary. The simple solution proposed in this paper is a boolean variable for each block, which indicates its completion. Initialised with `false`, the variable is set to `true` immediately after the block has finished. The dependent code block delays its execution until the variable becomes `true`. Of course, this has only to be done for those code blocks on which other blocks of different `section` constructs depend on.

The delay of dependent code blocks is implemented as an active wait with a loop spinning on the synchronisation variable. This is faster than suspending the thread, but can easily be changed to another mechanism, for example when it is desired to have idle waits in order to save energy.

Figure 3 displays the resulting code that is generated for the schedule of Figure 1. This is the same code as shown above, augmented with the necessary synchronisation variables. An example: as soon as code block B finishes is sets the corresponding variable `taskBDone` to `true`. The code blocks depending on B, i.e. E and G (see Figure 1), are delayed with `while{!taskBDone}{}` until `taskBDone` becomes true. Strictly, the `while{!taskBDone){}` before code block G is not necessary, but it simplifies the code generation. It will be easy to improve the code generation to remove such redundant waits in the future.

The resulting output code of the source-to-source complier only uses standard OpenMP directives and thus can be compiled with and standard OpenMP compiler. For the proof-of-concept implementation this was JOMP, the Java compiler for OpenMP-like directives [2,5].

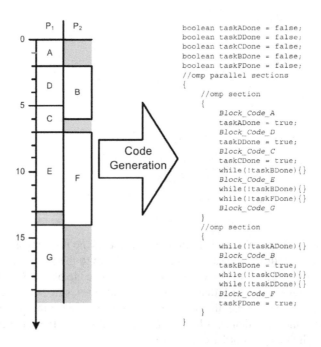

Fig. 3. Code generation from schedule

4   Practical Evaluation and Performance

Experiments have been conducted in order to evaluate the performance of the new tasks-task approach and the proof-of-concept complier. From the area of task scheduling the typical behaviour for different kinds of task graphs is well known, e.g. [6]. So the question that arises is whether the tasks-task approach and the complier implementation behave as predicted by those results. Further, one wants to evaluate the overhead that might be introduced with the new approach and its implementation.

4.1   Workload

For the workload of the experiments a large number of small programs was generated. Each program essentially consists of one tasks construct. Three parameters are varied for the construction of its code body:

1. Size/Number of task blocks – the tasks constructs contains $5, 6, 7, 10$ or 20 task blocks.
2. Density – the density is defined as the average number of dependences per task block. In terms of the corresponding task graph G, this is the number of edges divided by the number of nodes, $|E|/|V|$. There are five steps of density, taken from the range 0 to 4.95 for 20 task blocks, and 0 to ca. 2 for all other sizes.
3. Variance of the distribution of the weight values of the task blocks – The mean weight value is set to 1000 and there are three level of variance – low, medium and high – corresponding to a standard deviation of approximately 200, 500, 800 respectively.

About 60 different programs were generated by varying these three parameters. The actual code of a task block consisted of a small iterative kernel. To achieve a execution time of each task block that corresponds to its weight value, the number of iterations of the kernel was set accordingly.

In comparison to experimental evaluations of task scheduling algorithms, e.g. [6], the number of tasks is low. However, for the typical use case where a program part with task parallelism is parallelised manually, this number seems realistic. In contrast, the density is rather on the high side, but both choices will show the power of the proposed approach. Note that more task blocks and lower density are generally easier to parallelise.

4.2   Experimental Environment

The workload programs were compiled and executed on a Sun Fire v880, which is a centralised shared memory system with 8 processors. For the compilation the programs went through our source-to-source compiler, then JOMP and finally the Sun Java compiler. Each program was compiled and executed with 1-8 threads. To reduce measurement noise, each program was executed 10 times for each number of threads and the average value was recorded.

4.3 Results

In the following the major results are presented and discussed.

Figure 4 displays a chart of the speedup over the number of threads for programs with different sizes, i.e. number of `task` blocks. The density is fixed to 1 and the `weight` distribution variance is medium.

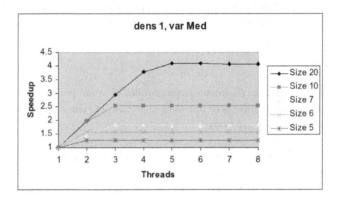

Fig. 4. Speedup over number of threads for programs of different sizes, density 1 and medium weight variance

All programs show the same qualitative behaviour: the speedup increases more or less linearly with the number of threads until a saturation point is reached. After this point, the speedup remains the same, even if more threads are used for the execution. Further, the speedup curve saturates later the more `task` blocks are in the program. For example, the program with 20 `task` blocks saturates with 5 threads, whereas the program with 10 `task` blocks saturates with 3 threads.

This is exactly the expected behaviour of such programs. The inherent concurrency is limited by the number of `task` blocks, but also by the dependences between them. With this in mind, very good speedups are achieved. For programs with more task blocks and similar dependence structure higher speedups can be expected.

Also interesting to observe is that after the saturation point the speedup remains stable and does not decrease. This is very desirable and demonstrates that no overhead is introduced for additional threads even if they cannot be employed efficiently. The scheduling algorithm achieves this automatically, because it leaves processors empty when their utilisation would decrease the performance. In turn the code generation does not create a `section` region for empty processors of the schedule, hence unnecessary threads are not involved in the execution of a `sections` construct.

Figure 5 displays a chart of the speedup over the number of threads for programs with different densities, number of dependences. The size is fixed to 20 `task` blocks and the `weight` distribution variance is medium.

For most programs the behaviour is similar as in Figure 4: the speedup increases linearly until a certain saturation point and then stays level. Also, for most programs it holds that the higher the density, the lower the speedup.

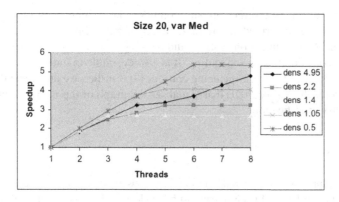

Fig. 5. Speedup over number of threads for programs of different densities, size 20 and medium variance

Two points are different though. First, the saturation point of the program with a density of 0.5 is with 6 threads higher than all saturation points in Figure 4. This makes perfect sense, as the density (0.5) is lower than the density (1) of the programs in Figure 4. In other words, the program has better inherent concurrency and can therefore be parallelised more efficiently. This demonstrates the general scalability of our approach.

Second, while the performance of the programs decreases with the increase of the density, there is one exception. The program with the highest density (4.95) has no saturation point and reaches for 8 threads almost the same speedup as the program with the lowest density (0.5). The reason lies in the dependence structure of the programs. Most have a pseudo-randomly created structure, but the program with density 4.95 is a highly regular program with two layers of nodes in the task graph. Nodes in the same layer are independent of each other, which implies a high inherent concurrency, hence it enables better scalability.

In summary, the experiments clearly demonstrated the successful parallelisation of the `tasks` constructs. The achieved scalability corresponds to the inherent parallelism of the programs and is in concordance with previous task scheduling results. Furthermore, the introduced overhead is low and does not jeopardise the performance gain for higher numbers of threads.

5   Conclusions

This paper presented an extension to OpenMP that allows the efficient parallelisation of fine grained tasks with an arbitrary dependence structure. The new `tasks-task` directives are easy to employ and generalise the `sections-section` concept. Parallelising a program with these constructs maintains the sequential semantic and integrates smoothly into the normal parallelisation process. A proof-of-concept source-to-source complier was developed and the approach was evaluated in experiments. The results show that the new directives are capable of exploiting the inherent task parallelism in the code without introducing any additional overhead.

In the future, the new directives and compiler can be enhanced in several ways. Scheduling can be performed during runtime, making the approach even more flexible and adaptable. Communication costs should be integrated into the `tasks` construct to permit their consideration in scheduling. It is also desirable to unify the new construct with the `taskq-task` work-queueing concept [8]. Further, we are currently working on a plug-in for the Eclipse IDE that visualises the graph of the `tasks` construct and allows its easy modification.

References

1. OpenMP Application Program Interface, http://www.openmp.org/specs/
2. Bull, J.M., Westhead, M.D., Kambites, M.E., Obdržálek, J.: Towards OpenMP for Java. In: Proc. of 2nd European Workshop on OpenMP, Edinburgh, UK, September 2000, pp. 98–105 (2000)
3. Garey, M.R., Johnson, D.S.: Computers and Intractability: A Guide to the Theory of NP-Completeness. Freeman (1979)
4. Hu, T.: Parallel sequencing and assembly line problems. Operations Research 9(6), 841–848 (1961)
5. Kambites, M.E., Obdržálek, J., Bull, J.M.: An OpenMP-like interface for parallel programming in Java. Concurrency and Computation: Practice and Experience 13(8-9), 793–814 (2001)
6. Kwok, Y.-K., Ahmad, I.: Benchmarking the task graph scheduling algorithms. In: Proc. of Int. Par. Processing Symposium/Symposium on Par. and Distributed Processing (IPPS/SPDP 1998), Orlando, Florida, USA, April 1998, pp. 531–537 (1998)
7. Kwok, Y.-K., Ahmad, I.: A comparison of parallel search-based algorithms for multiprocessors scheduling. In: Proc. of the 2nd European Conference on Parallel and Distributed Systems (EUROPDS 1998), Vienna, Austria (July 1998)
8. Shah, S., Haab, G., Petersen, P., Throop, J.: Flexible control structures for parallelism in OpenMP. Concurrency: Practice and Experience 12(12), 1219–1239 (2000); Special Issue: EWOMP 1999 - First European Workshop on OpenMP
9. Sinnen, O.: Task Scheduling for Parallel Systems. Wiley, Chichester (2007)
10. Sinnen, O., Kozlov, A.V., Semar Shahul, A.Z.: Optimal scheduling of task graphs on parallel systems. In: Proc. Int. Conference on Parallel and Distributed Computing and Networks, Innsbruck, Austria (February 2007)
11. Su, E., Tian, X., Girkar, M., Haab, G., Shah, S., Petersen, P.: Compiler support for workqueuing execution model for Intel SMP architectures. In: Proc. of European Workshop on OpenMP (EWOMP) (September 2002)

Performance Evaluation of a Multi-zone Application in Different OpenMP Approaches

Haoqiang Jin[1], Barbara Chapman[2], and Lei Huang[2]

[1] NAS Division, NASA Ames Research Center,
Moffett Field, CA 94035-1000
hjin@nas.nasa.gov
[2] Department of Computer Science, University of Houston,
Houston, TX 77004
{chapman,leihuang}@cs.uh.edu

Abstract. We describe a performance study of a multi-zone application benchmark implemented in several OpenMP approaches that exploit multi-level parallelism and deal with unbalanced workload. The multi-zone application was derived from the well-known NAS Parallel Benchmarks (NPB) suite that involves flow solvers on collections of loosely coupled discretization meshes. Parallel versions of this application have been developed using the Subteam concept and Workqueuing model as extensions to the current OpenMP. We examine the performance impact of these extensions to OpenMP on a large shared memory machine and compare with hybrid and nested OpenMP programming models.

1   Introduction

Since its introduction in 1997, OpenMP has become the de facto standard for shared memory parallel programming. The notable advantages of the model are its global view of memory space that simplifies programming development and its incremental approach toward parallelization. However, it is a big challenge to scale OpenMP codes to tens or hundreds of processors. One of the difficulties is a result of limited parallelism that can be exploited on a single level of loop nest. Although the current standard [8] allows one to use nested OpenMP parallel regions, the performance is not very satisfactory. One of the known issues with nested OpenMP is its lack of support for thread team reuse at the nesting level, which affects the overall application performance and will be more profound on multi-core, multi-chip architectures. There is no guarantee that the same OS threads will be used at each invocation of parallel regions although many OS and compilers have provided support for thread affinity at a single level. To remedy this deficiency, the NANOS compiler team [1] has introduced the GROUPS clause to the outer parallel region to specify a thread group composition prior to the start of nested parallel regions, and Zhang [13] proposed extensions for thread mapping and grouping.

Chapman and collaborators [5] proposed the *Subteam* concept to improve work distribution by introducing subteams of threads within a single level of

B. Chapman et al. (Eds.): IWOMP 2007, LNCS 4935, pp. 25–36, 2008.

thread team, as an alternative for nested OpenMP. Conceptually, a subteam is similar to a process subgroup in the MPI context. The user has control over how threads are subdivided in order to suit application needs. The subteam proposal introduced an `onthreads` clause to a work-sharing directive so that the work will be performed among the subset of threads, including the implicit barrier at the end of the construct.

One of the prominent extensions to the current OpenMP is the *Workqueuing* (or Taskq) model first introduced by Shah et al. [9] and implemented in the Intel C++ compiler [10]. It was designed to work with recursive algorithms and cases where work units can only be determined dynamically. Because of its dynamic nature, Taskq can also be used effectively in an unbalanced workload environment. The Taskq model will be included in the coming OpenMP 3.0 release [4]. Although the final *tasking* directive in OpenMP 3.0 might not be the same as the original Intel Taskq proposal, it should still be quite intuitive to understand what potential the more dynamic approach can offer to applications.

In this study, we will compare different OpenMP approaches for the parallelization of a multi-zone application benchmark on a large shared memory machine. In section 2, we briefly discuss the application under consideration. The different implementations of our benchmark code are described in section 3 and the machine description and performance results are presented in section 4. We conclude our study in section 5 where we also elaborate on future work.

2   Multi-zone Application Benchmark

The multi-zone application benchmarks were developed [6,11] as an extension to the original NAS Parallel Benchmarks (NPBs) [2]. These benchmarks involve solving the application benchmarks BT, SP, and LU on collections of loosely coupled discretization meshes (or zones). The solutions on the meshes are updated independently, but after each time step they exchange boundary value information. This strategy, which is common among many production structured-mesh flow solver codes, provides relatively easy to exploit coarse-grain parallelism between zones. Since the individual application benchmark also allows fine-grain parallelism within each zone, this NPB extension, named NPB Multi-Zone (NPB-MZ), is a good candidate for testing hybrid and multi-level parallelization tools and strategies.

NPB-MZ contains three application benchmarks: BT-MZ, SP-MZ, and LU-MZ, with problem sizes defined from Class S to Class F. The difference between classes comes from how the number of zones and the size of each zone are defined in each benchmark. We focus our study on the BT-MZ benchmark because it was designed to have uneven-sized zones, which allows us to test various load balancing strategies. For example, the Class B problem has 64 zones with sizes ranging from 3K to 60K mesh points. Previously, the hybrid MPI+OpenMP [6] and nested OpenMP [1] programming models have been used to exploit parallelism in NPB-MZ beyond a single level. These approaches will be briefly described in the next section.

3 Benchmark Implementations

In this section, we describe five approaches of using OpenMP or its extension to implement the multi-zone BT-MZ benchmark. Three of the approaches exploit multi-level parallelism and the other two are concerned with balancing workload dynamically.

3.1 Hybrid MPI+OpenMP

The hybrid MPI+OpenMP implementation exploits two levels of parallelism in the multi-zone benchmark in which OpenMP is applied for fine grained intra-zone parallelization and MPI is used for coarse grained inter-zone parallelization. Load balancing in BT-MZ is based on a bin-packing algorithm with an additional adjustment from OpenMP threads [6]. In this strategy, multiple zones are clustered into zone groups among which the computational workload is evenly distributed. Each zone group is then assigned to an MPI process for parallel execution. This process involves sorting zones by size in descending order and bin-packing into zone groups. Exchanging boundary data within each time step requires MPI many-to-many communication. The hybrid version is fully described in Ref. [6] and is part of the standard NPB distribution. We will use the hybrid version as the baseline for comparison with other OpenMP implementations.

3.2 Nested OpenMP

The nested OpenMP implementation is based on the two-level approach of the hybrid version except that OpenMP is used for both levels of parallelism. The inner level parallelization for loop parallelism within each zone is essentially the same as that of the hybrid version. The only addition is the "num_threads" clause to each inner parallel region to specify the number of threads. The first (outer) level OpenMP exploits coarse-grained parallelism between zones.

A sketch of the iteration loop using the nested OpenMP code is illustrated in Fig. 1. The outer level parallelization is adopted from the MPI approach: workloads from zones are explicitly distributed among the outer-level threads. The difference is that OpenMP now works on the shared data space as opposed to private data in the MPI version. The load balancing is done statically through the same bin-packing algorithm where zones are first sorted by size, then assigned to the least loaded thread one by one. The routine "map_zones" returns the number and list of zones (num_proc_zones and proc_zone_id) assigned to a given thread (myid) as well as the number of threads (nthreads) for the inner parallel regions. This information is then passed to the "num_threads" clause in the solver routines. The MPI communication calls inside "exch_qbc" for boundary data exchange are replaced with direct data copy and proper barrier synchronization.

In order to reduce the fork-and-join overhead associated with the inner-level parallel regions, a variant was also created: a single parallel construct is applied to the time step loop block and all inner-parallel regions are replaced with orphaned "omp do" constructs. This version, namely, version 2, will be discussed together with the first version in the results section.

```
!$omp parallel private(myid,...)              !$omp parallel private(myid,...)
  myid = omp_get_thread_num()                   myid = omp_get_thread_num()
  call map_zones(myid,..,nthreads,&             call map_zones(myid,..,mytid,&
          & num_proc_zones,proc_zone_id)                & proc_thread_team)
  do step=1,niter                               t1 = proc_thread_team(1,mytid)
    call exch_qbc(u,...,nthreads)               t2 = proc_thread_team(2,mytid)
    do iz = 1, num_proc_zones                   do step=1,niter
      zone = proc_zone_id(iz)                     call exch_qbc(u,...,t1,t2)
      call adi(u(zone),..,nthreads)              do iz = 1, num_proc_zones
    end do                                          zone = proc_zone_id(iz)
  end do                                            call adi(u(zone),..,t1,t2)
!$omp end parallel                                end do
                                                end do
  subroutine adi(u,..,nthreads)               !$omp end parallel
!$omp parallel do &
!$omp& num_threads(nthreads)                    subroutine adi(u,..,t1,t2)
  do k=2,nz-1                                  !$omp do onthreads(t1:t2:1)
    solve for u in the current zone             do k=2,nz-1
  end do                                          solve for u in the current zone
                                                end do
```

Fig. 1. Sample nested OpenMP code on the left and Subteam code on the right

3.3   Subteam in OpenMP

The subteam version was derived from the nested OpenMP version. Changes include replacing the inner level parallel regions with orphaned "omp do" constructs and adding the "onthreads" clause to specify the subteam composition. The sample subteam code is listed in the right panel of Fig. 1. The main difference is in the call to "map_zones." This routine determines which subteam (mytid) the current thread belongs to and what members are in the current subteam (proc_thread_team). This process could be simplified by introducing a runtime function for subteam formation and management, which is not included in the subteam proposal [5].

The same load-balancing scheme as described in previous sections is applied in the subteam version to create zone groups. Each subteam works on one zone group, and thus, the number of zone groups equals the number of subteams. We use an environment variable to specify the number of subteams at runtime. Threads assigned to each subteam will work on loop-level parallelism within each zone. There is no overlapping of thread ids among different subteams. Similar to the nested OpenMP version, the routine "exch_qbc" uses direct array copy and proper global barrier synchronization for boundary communication.

3.4   OpenMP at Outer Level

One of the advantages of OpenMP is its ability to handle unbalanced workload in a dynamic fashion without much user intervention. The programming effort is much less than the explicit approach described in previous sections for handling load balance. The tradeoff is potentially higher overhead associated with dynamic scheduling and less thread-data affinity as would be achieved in a static approach. To examine the potential performance tradeoff, we developed an OpenMP version that solely focuses on the coarse-grained parallelization of different zones of the multi-zone benchmark. As illustrated in Fig. 2 left panel, this version is much simpler and compact. The "omp do" directive is applied to the loop nest over multiple zones. There is no explicit coding for load balancing, which is achieved through the OpenMP dynamic runtime schedule. The use of the "schedule(runtime)" clause allows us to compare different OpenMP loop schedules. A "zone_sort_id" array is used to store zone ids in different sorting schemes.

```
!$omp parallel private(zone,...)          #pragma omp parallel private(zone)
 do step=1,niter                           for (step=1;step<=niter;step++) {
   call exch_qbc(u,...)                       exch_qbc(u,...);
   !$omp do schedule(runtime)                 #pragma intel omp taskq
   do iz = 1, num_zones                       for (iz=0;iz<num_zones;iz++) {
     zone = zone_sort_id(iz)                    zone = zone_sort_id[iz];
     call adi(u(zone),...)                      #pragma intel omp task \
   end do                                                   captureprivate(zone)
 end do                                           adi(&u[zone],...);
!$omp end parallel                        }}
```

Fig. 2. Code segment using OpenMP runtime scheduling (left) and Intel `taskq` directives (right)

3.5   Workqueuing Model

We developed a Taskq version of the BT-MZ benchmark based on the Intel workqueuing model. Because Intel implemented Taskq only in its C++ compiler for C/C++ applications and there is no other vendor compiler available for testing the concept, we had to first convert the Fortran implementation of BT-MZ to the C counterpart. To minimize the performance impact from such a conversion, we did the following:

- Fortran multi-dimensional arrays are converted to linearized C arrays, such as

  ```
  u(m,i,j,k) -> u[m+5*(i+nxmax*(j+ny*k))],
  ```

- The `restrict` qualifier is added to pointer variables in subroutine argument to enable compiler to perform optimization without pre-assumed pointer aliasing, for example

  ```
  void add(double *restrict u, double *restrict rhs,..).
  ```

Once we had the C version, the Taskq implementation of the BT-MZ benchmark (Fig. 2 right panel) is straightforward. Each work unit for a task is defined by the solver on an individual zone. The "intel omp taskq" directive is added to the loop nest over zones. Inside the zone loop nest, the "intel omp task" directive is used to generate tasks for each loop iteration and the *zone* value is preserved for each task by the "captureprivate" clause. The implicit synchronization at the end of the taskq construct guarantees the completion of all tasks before going to the next iteration. Again, to test the performance impact of workload ordering, we use "zone_sort_id" to store the sorted zone ids.

4   Performance Results

In this section, we present performance results obtained on a large shared memory system. We will first give a brief description of the system and programming support.

4.1   Testing Environment

For our study, we used an SGI Altix 3700BX2 system that is one of the 20 nodes in the Columbia supercomputer installed at NASA Ames Research Center [3]. The Altix BX2 node has 512 Intel Itanium 2 processors, each clocked at 1.6 GHz and containing 9 MB on-chip L3 data cache. Approximately 1 TB of global shared-access memory is provided through the SGI scalable non-uniform memory access flexible (NUMAflex) architecture. The underlying NUMAlink4 interconnect provides 6.4 GB/s bandwidth and scales linearly with the number of processors. A single Linux operating system runs on the Altix system, providing an ideal environment for shared-memory programming such as OpenMP.

The system is equipped with SGI message-passing toolkit (MPT 1.12) that supports MPI programming. We used the Intel Fortran, C/C++ 9.1 compilers for IA64 that support OpenMP 2.5 as well as the Taskq model. All of our experiments were run under the PBSpro batch system in a shared environment. In order to reduce variation in timing and improve performance, the "dplace" placement tool was used to bind processes/threads to physical processors.

For testing the OpenMP Subteam concept as described in Section 3.3, we employed the Open64 research compiler [7] that was extended to support the "onthreads" clause. This is essentially a source-to-source translation process and the generated code is then compiled with a native compiler. A small runtime library was developed to support basic subteam functions, such as loop iteration scheduling and synchronization for subteam threads.

4.2   Multi-level Parallelism

In order to compare different multi-level parallel versions of the BT-MZ benchmark, we first examine the performance impact from varying the number of zone groups on a given number of CPUs. The left panel of Fig. 3 plots benchmark timing in seconds as a function of the number of groups at 32 CPUs for the Class B problem size. The notation "$N_g \times N_t$" denotes the number of zone groups (N_g) formed for the first level parallelism and the number of threads (N_t) for the second level parallelism within each group. In the hybrid MPI+OpenMP version, N_g is the same as the number of MPI processes and N_t is the number of OpenMP threads per MPI process. N_g in the nested OpenMP versions is the number of outer-level threads, and in the subteam version is the number of subteams.

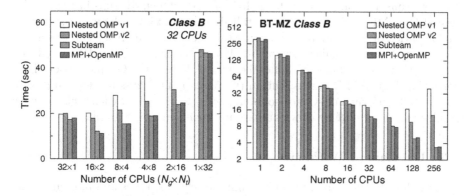

Fig. 3. Timing comparison of nested OpenMP, Subteam, and MPI+OpenMP versions of BT-MZ for the Class B problem, on the left for a given number of CPUs and on the right as a function of CPU counts

Overall the subteam version is very close in performance to the MPI+OpenMP hybrid version. This indicates that the data layout of the subteam version is very similar to that of the hybrid version, even though Subteam uses shared data arrays and MPI uses private data arrays. At single level parallelization, either $N \times 1$ or $1 \times N$, the performance of three approaches is very close. Between the two ends, the nested-OpenMP v1 performs consistently 30-80% worse than the other two versions. By reducing the number of inner-level parallel regions in the second version (v2), the performance of nested OpenMP improved substantially, although it still lags behind. The large overhead associated with the inner parallel regions is likely due to the inability of the OpenMP runtime library to reuse threads efficiently at the second level. Even though the *dplace* tool binds the first-level threads properly, it has no control over the second-level threads. This result is consistent with the previous observation by Ayguade et al. [1]

The best performance is achieved by maximizing the number of zone groups as long as the workload can be balanced. For Class B, the optimal number of zone groups is 16. Beyond 16 CPUs, multi-level parallelism is needed for additional performance gain. Both subteam and hybrid versions follow this analysis, but the nested OpenMP tends to prefer a larger number of threads at the outer level, especially when the total number of CPUs increases.

The scaling results of BT-MZ from the best combinations of zone-groups and threads are summarized in right panel of Fig. 3. Both the subteam and hybrid versions scale well up to the measured CPU counts. Up to 16 CPUs, when only the outer-level parallelism is employed, the nested OpenMP versions performs similarly to the other two versions. Beyond 16 CPUs, nested OpenMP suffers from large overhead associated with the second-level parallelism and becomes much worse at larger CPU counts.

To understand better why the nested OpenMP codes suffer from performance degradation in the multi-level mode, we collected additional performance information from hardware counters available on the Altix and the results from the 8×4 runs are compared with the hybrid MPI+OpenMP runs in Fig. 4. The nested OpenMP v1 has the highest stalled cycles and L3 cache misses, which is an indication of thread-data mismatch. Stalled cycle is usually a result of waiting on resources, in particular memory. Although the nested OpenMP v2 reduced stalled cycles, but it has large L3 cache misses. Three pure OpenMP codes have somewhat higher TLB misses; but on the Altix, a TLB miss has less impact on the overall performance. Other counters, such as L1 and L2 cache misses, have similar values for all four codes and are not included in the graph.

4.3   Unbalanced Workload

To test the effectiveness of OpenMP runtime schedule kinds and more dynamic approaches on unbalanced workload, we focus on the single-level OpenMP versions of BT-MZ as described in Sections 3.4 and 3.5, which exploit parallelism among unbalanced zones. No nested parallelism is considered here.

4.3.1 Impact of Schedule Kind
The results from 16-thread runs using different runtime schedule kinds and chunk sizes are shown in Fig. 5. The "`dynamic,1`" schedule produces the best result for the given problem. As the chunk size increases, the performance decreases. The

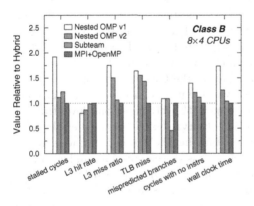

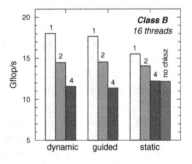

Fig. 4. Comparison of hardware performance counter results obtained on the Altix for the 8×4 runs of the four BT-MZ versions

Fig. 5. Performance comparison of different schedule kinds for BT-MZ Class B, 16 threads. Numbers in the graph indicate chunk sizes. The last bar is for a `static` schedule without chunk size.

"guided" schedule is only slightly worse. The "static" schedule without chunk size (the last bar in the graph) shows its limitation in dealing with unbalanced workload and is as much as 50% worse than the "dynamic,1" schedule. The "static,1" (or cyclic) schedule improves the performance but not sufficiently.

4.3.2 Workload Ordering on Performance

As noted in the benchmark description, the zone workload in BT-MZ was designed to be uneven. Class B contains 64 zones whose sizes, shown in Fig. 6 on the left, range from 3K to 60K mesh points. The right graph in Fig. 6 shows the performance impact of three different orderings of zones in size on the "dynamic,1" schedule: natural (original) order, descending order, and ascending order. For comparison, the graph also includes results from a single-level OpenMP version that uses the static bin-packing algorithm for load balancing. This version is essentially the same as the nested OpenMP v1 described in Section 3.2 without the nested parallelism. We observe that by sorting zones into descending order, the performance can improve by as much as 45% (18 to 26 Gflop/s on 16 threads). This result supports the observation reported by Van Zee et al. [12] in their FLAME code using the workqueuing model.

The impact of different workload orderings on the "guided" schedule (not shown in the graph) is very similar to that on the "dynamic" schedule. It is worth noting that the dynamic approach for unbalanced workload is only slightly worse (~15% beyond 16 threads) than the static bin-packing approach. However, the programming effort in the former case is considerably less.

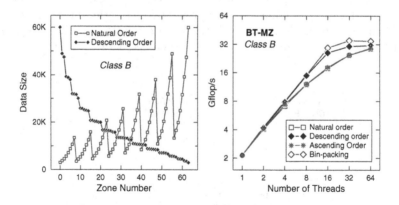

Fig. 6. Performance impact of different workload orderings on the "dynamic" schedule. Results from the static bin-packing approach are included for comparison.

4.3.3 Workqueuing Model

Before going into the workqueuing (or taskq) model, we first examine the performance change as a result of converting the code from Fortran to C. Due to pointer aliasing, a C code can suffer from the constraint in compiler optimization for pointers. In order to reduce or even eliminate pointer aliasing, one can either use the "`restrict`" modifier or rely on compiler flags. The Intel compiler provides the option "`-fno-alias`" for this purpose. Table 1 summarizes the results of the OpenMP C version of BT-MZ using different compiler aliasing options and compares with the Fortran version. The no-alias option produces more than twice as much improvement in performance as the default aliasing option. Combining with the "`restrict`" modifier, the C code performs very close to the Fortran counterpart. This combined option was used in collecting the C results below.

Table 1. Comparison of results from different aliasing options for the Class B, BT-MZ on 16 threads

Case		Time(sec)	Gflop/s	Compiler option
	Default aliasing	49.86	12.059	
	"restrict" keyword	37.17	16.175	`-restrict`
C	No aliasing	23.49	25.594	`-fno-alias`
	Combination	23.38	25.718	`-restrict -fno-alias`
Fortran		23.01	26.124	

Figure 7 compares the Intel Taskq version of BT-MZ with the single-level OpenMP versions (both C and Fortran) using dynamic scheduling for load balancing. It is encouraging to note that the Taskq version has similar performance to the single-level OpenMP C version using the "`dynamic,1`" schedule up to 32 threads. Only at 64 threads the dynamic-schedule version outperforms the Taskq version by about 20%. As illustrated by the two panels in the figure, sorting workload into descending order improves overall performance for Taskq as well. Comparing to the Fortran version, the performance of Taskq gets worse at larger thread counts, primarily due to the difference between Fortran and C.

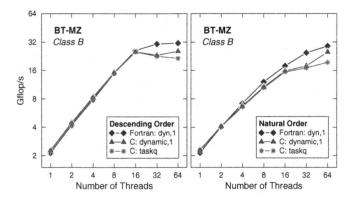

Fig. 7. Performance comparison of the Taskq version with the single-level OpenMP versions (in both C and Fortran) using the "`dynamic,1`" schedule

5   Conclusion

We have presented performance evaluation of four different OpenMP approaches in dealing with multi-level parallelism and unbalanced workload, and compared with a hybrid MPI+OpenMP method. The nested OpenMP approach suffered from performance degradation as a result of large overhead and lack of thread reuse when invoking the inner level parallelism. By minimizing the number of inner level parallel regions we improved nested OpenMP performance dramatically. Another potential way to reduce overhead associated with nested parallel regions is by predefining a thread tree structure as proposed in [1] and [13] so that the runtime can perform better scheduling optimization.

The approach based on the Subteam extension to OpenMP overcame some of the limitations with nested OpenMP and showed promise in achieving performance close to that of the hybrid MPI+OpenMP method. Our study also points out the importance of extending the Subteam proposal to include API for subteam creation and management.

It is very encouraging that the more dynamic approach provided by the workqueuing model showed great potential in dealing with unbalanced workload. This model can benefit from using a weight factor in scheduling tasks.

For future work, we would like to conduct our experiments on more platforms, in particular to study the support of nested parallelism from different compilers and runtime systems. A natural extension is to investigate the performance characteristics of nested parallelism under the workqueuing model. It is also important to extend our experience from a single benchmark application to more realistic applications.

Acknowledgements

The authors would like to acknowledge fruitful discussions with Robert Hood, Johnny Chang, and support from the staff at NAS division for many experiments conducted on the Columbia supercomputer.

References

1. Ayguade, E., Gonzalez, M., Martorell, X., Jost, G.: Employing Nested OpenMP for the Parallelization of Multi-Zone Computational Fluid Dynamics Applications. Monien, B. (ed.) J. of Parallel and Distributed Computing, special issue, 66(5), 686 (2006)
2. Bailey, D., Barton, J., Lasinksi, T., Simon, H.: The NAS Parallel Benchmarks. NAS Technical Report RNR-91-002, NASA Ames Research Center (1991)
3. Biswas, R., Djomehri, M.J., Hood, R., Jin, H., Kiris, C., Saini, S.: An Application-Based Performance Characterization of the Columbia Supercluster. In: Proc. of the ACM/IEEE SC 2005 Conference (2005)
4. Bull, M.: OpenMP 3.0 Overview. In: OpenMP BoF at the SC 2006 conference (2006), http://www.compunity.org/futures/

5. Chapman, B., Huang, L., Jin, H., Jost, G., de Supinski, B.: Toward Enhancing OpenMP's Work-Sharing Directives. In: Nagel, W.E., Walter, W.V., Lehner, W. (eds.) Euro-Par 2006. LNCS, vol. 4128, pp. 645–654. Springer, Heidelberg (2006)
6. Jin, H., Van der Wijngaart, R.F.: Performance Characteristics of the Multi-Zone NAS Parallel Benchmarks. Monien, B. (ed.) J. of Parallel and Distributed Computing, special issue, 66(5), 674 (2006)
7. Open64 Research Compiler, http://www.open64.net/
8. The OpenMP Standard, http://www.openmp.org/
9. Shah, S., Haab, G., Petersen, P., Throop, J.: Flexible Control Structure for Parallelism in OpenMP. In: European Workshop on OpenMP (EWOMP 1999) (1999)
10. Su, E., Tian, X., Girkar, M., Haab, G., Shah, S., Petersen, P.: Compiler Support of the Workqueuing Execution Model for Intel SMP Architectures. In: European Workshop on OpenMP (EWOMP 2002) (2002)
11. Van der Wijngaart, R.F., Jin, H.: The NAS Parallel Benchmarks, Multi-Zone Versions. NAS Technical Report NAS-03-010, NASA Ames Research Center (2003), http://www.nas.nasa.gov/Software/NPB/
12. Van Zee, F., Bientinesi, P., Low, T.M., Van de Geijn, R.: Scalable Parallelization of FLAME Code via the Workquenuing Model. ACM Trans. on Math.Software (submitted, 2006)
13. Zhang, G.: Extending the OpenMP Standard for Thread Mapping and Grouping. In: International Workshop on OpenMP (IWOMP 2006), Reims, France (2006)

Transactional Memory and OpenMP

Miloš Milovanović, Roger Ferrer, Osman S. Unsal, Adrian Cristal,
Xavier Martorell, Eduard Ayguadé, Jesús Labarta, and Mateo Valero

Barcelona Supercomputing Center
c/ Jordi Girona 31, 08034 Barcelona, Spain
{milos.milovanovic,roger.ferrer,osman.unsal,adrian.cristal,
xavier.martorell,eduard.ayguade,jesus.labarta,mateo.valero}@bsc.es
http://www.bsc.es

Abstract. Future generations of Chip Multiprocessors (CMP) will provide dozens or even hundreds of cores inside the chip. Writing applications that benefit from the massive computational power offered by
these chips is not going to be an easy task for mainstream programmers who are used to sequential algorithms rather than parallel ones.
This paper explores the possibility of using Transactional Memory (TM)
in OpenMP, the industrial standard for writing parallel programs on
shared-memory architectures, for C, C++, and Fortran. One of the major complexities in writing OpenMP applications is the use of critical
regions (locks), atomic regions and barriers to synchronize the execution of parallel activities in threads. TM has been proposed as a mechanism that abstracts some of the complexities associated with concurrent
access to shared data while enabling scalable performance. The paper
presents a first proof-of-concept implementation of OpenMP with TM.
Some extensions to the language are proposed to express transactions.
These extensions are handled by our source-to-source OpenMP Mercurium compiler and our Software Transactional Memory (STM) library
Nebelung that supports the code generated by Mercurium. The current
implementation of the library has no support at the hardware level, so it
is a proof-of-concept implementation. Hardware Transactional Memory
(HTM) or Hardware-assisted STM (HaSTM) are seen as possible paths
to make the tandem TM-OpenMP more usable. The paper finishes with
a set of open issues that still need to be addressed, either in OpenMP or
in the hardware/software implementations of TM.

Keywords: Compiler, OpenMP, Software Transaction Memory, STM
Library.

1 Introduction

The trend towards incorporating more cores in Chip Multiprocessors (CMP)
will continue, with the potential for hundreds of cores for future technology
generations. Inefficient data access ordering and synchronization primitives such
as locking will limit programmer productivity and application performance for

B. Chapman et al. (Eds.): IWOMP 2007, LNCS 4935, pp. 37–53, 2008.

those future processors. Transactional Memory (TM) is a crucial mechanism to tackle this problem by abstracting away the complexities associated by concurrent access to shared data [1] where multiple threads need to simultaneously access shared memory locations atomically.

OpenMP [2], the industrial standard for writing parallel programs on shared-memory architectures, for C, C++, and Fortran is a "traditional" programming model in terms of mechanisms offered to guarantee mutual exclusion. It offers a set of low-level primitives around locks and the high-level `critical` construct to protect the access to shared data (through the ownership of one or more locks). Lock-based mechanisms are complex to use and error prone especially when trying to avoid deadlock situations or when trying to use fine-grain locking to achieve better scalability on highly parallel hardware. Consequently there is currently concern in the programming and computer architecture communities that a parallel programming productivity/performance wall might be looming in the horizon.

Transactional Memory (TM) is a promising mechanism to tackle this problem by abstracting some of the complexities associated with concurrent access to shared data. With TM, multiple threads can simultaneously try to access multiple shared memory locations within the scope of what is called a "transaction". The detection of memory access conflicts causes transactions to rollback.

When compared to TM, locks are pessimistic. With mutual exclusion, only one thread can hold a given lock at a given time whereas with TM more than one thread can access a given critical section simultaneously. Given that actual conflicts are rare in many programs [3], the optimistic TM approach makes much more sense as a future programming model. With TM the programmer specifies intent rather than mechanism (i.e. the programmer can focus on determining where atomicity is necessary, rather than the mechanisms used to enforce it), resulting in a higher-level abstraction than locks.

Figure 1 shows an excerpt from the SpecOMP2001 AMMP program. In this case the programmer uses a lock for each atom (`a1->lock` and `a2->lock`) to protect the access to the fields of the two interacting atoms (`a1` and `a2`). And this is the lock that it is used in each critical region. Due to nature of the program, it is very rare that more than one processor tries to access the critical region with the same lock.

Some recently proposed programming models, such as Sun's Fortress [4], IBMs X10 [5] and Crays Chapel [6], include an atomic statement to define conditional and/or conditional atomic blocks of statements that are executed as transactions. In some cases, atomic can also be an attribute for variables so that any update to them in the code is treated as if the update is in a short atomic section. OpenMP also offers an `atomic` pragma to specify certain indivisible read-operation-write sequences, for which current microprocessor usually provide hardware support. For example through load linked-store conditional (LL-SC).

Two main TM implementation styles stand out: hardware- and software-based. Historically, the earliest design proposals were hardware based. Software Transactional Memory (STM) [7] has been proposed to address, among other

```
#pragma omp parallel default(none)
     shared(dielectric, lambda, a_number, atomall)
     private (imax, i, a1, a2, xt, yt, zt, fx, fy, fz, a1fx, a1fy,
               a1fz, ii, jj, ux, uy, uz, r, k, r0)
{ ... #pragma omp for schedule(guided)
   for( i= 0; i< imax; i++) {
       ...
       a1 = (*atomall)[i];
       ...
       for( ii=0; ii< jj;ii++) {
           a2 = a1->close[ii];
#ifdef _OPENMP
           omp_set_lock(&(a2->lock));
#endif
           ux = (a2->dx -a1->dx)*lambda + (a2->x -a1->x);
           ...
           r =one/( ux*ux + uy*uy + uz*uz);
           r0 = sqrt(r);
           ux = ux*r0;
           ...
           k = -dielectric*a1->q*a2->q*r;
           r = r*r*r;
           k = k + a1->a*a2->a*r*r0*six;
           k = k - a1->b*a2->b*r*r*r0*twelve;
           a1fx = a1fx + ux*k;
           ...
           a2->fx = a2->fx - ux*k;
           ...
#ifdef _OPENMP
           omp_unset_lock(&(a2->lock));
#endif
       }
#ifdef _OPENMP
       omp_set_lock(&(a1->lock));
#endif
       a1->fx += a1fx ;
       a1->fy += a1fy ;
       a1->fz += a1fz ;
#ifdef _OPENMP
       omp_unset_lock(&(a1->lock));
#endif
   }
} /* omp parallel pragma */
```

Fig. 1. Excerpt from the SpecOMP2001 AMMP program

things, some inherent limitations of earlier forms of Hardware Transactional Memory (HTM) [8] such as a lack of commodity hardware with the proposed features, or a limitation to the number of locations that a transaction can access.

Beyond these two main approaches, two additional mixed approaches have recently been considered. Hybrid Transactional Memory (HyTM) [9], [10] supports transactional execution that generally occurs using HTM transaction but which backs off to STM transactions when hardware resources are exceeded. Hardware-assisted STM (HaSTM) combines STM with new architectural support to accelerate parts of the STMs implementation [11], [12]. These designs are both active research topics and provide very different performance characteristics: HyTM provides near-HTM performance for short transactions, but a "performance cliff" when falling back to STM. In contrast, HaSTM may provide performance some way between HTM and STM.

2   Basic Concepts

A transaction is a sequence of instructions, including reads and writes to memory, that either executes completely (commit) or has no effect (abort). When a transaction commits, all its writes are made visible and values can be used by other transactions. When a transaction is aborted, all its speculative writes are discarded.

Commit or abort are decided based on the detection of memory conflicts among parallel transactions. In order to detect and handle these conflicts, each running transaction is typically associated with a 'read set' and a 'write set'. Inside a transaction, the execution of each transactional memory read instruction adds the memory address to the read set. Each transactional memory write instruction adds the memory address and value to the write set of the transaction.

Conflict detection can be either eager or lazy. Eager conflict detection checks every individual read and write to see if there is a conflicting operation in another transaction. Eager conflict detection requires that the read and write sets of a transaction are visible to all the other transactions in the system. On the other hand, with lazy conflict detection a transaction waits until it tries to commit before checking its read and write sets against the write sets of other transactions.

Another fundamental design choice is how to resolve a conflict once it has been detected. Usually, if there is a conflict it is necessary to resolve it by immediately aborting one of the transactions involved in the conflict.

In order to support the execution of a transaction, a data versioning mechanism is needed to record the speculative writes. This speculative state should be discarded on an abort or used to update the global state on a successful commit. The two usual approaches to implement data versioning are based on using an undo-log or using buffered updates. Using an undo-log, a transaction applies updates directly to memory locations while logging the necessary information to undo the updates in case of abort. On the contrary, approaches using buffered updates keep the speculative state in a transaction-private buffer until commit time; if the commit succeeds, the original values before the store instructions are dropped and the speculative stores of the transaction are committed to memory.

HTM proposed systems keep the speculative state of the transactions mostly in the data cache or in a hardware buffer area. Transactional loads and stores can be kept in a separate "transactional cache" or in the conventional data caches, augmented with transactional support. In both cases the modifications are minimal since transactional support relies on extending existing cache coherence protocols such as MESI to detect conflicts and enforce atomicity. On the contrary, STM implementations must provide mechanisms for concurrent transactions to maintain their own views of the heap, allowing a transaction to see its own writes as it continues to run, and allowing memory updates to be discarded if the transaction ultimately aborts. In addition, STM implementations must provide mechanisms for detecting and resolving conflicts between transactions.

3   Our "Proof-of-Concept" Approach

In order to explore how TM could influence the future design and implementation of OpenMP, we have adopted a "proof-of-concept" approach based on a source-to-source code restructuring process, implemented in Mercurium [13], and two libraries to support the OpenMP execution model and to support transactional memory: NthLib [14] and Nebelung [15], respectively. We also propose some OpenMP extensions to specify transactions.

3.1   Is OpenMP ATOMIC a Transaction?

The `atomic` construct in OpenMP ensures that a specific storage location is updated atomically, rather than exposing it to the possibility of multiple, simultaneous writing threads.

```
#pragma omp atomic
   expression-statement
```

where expression-statement can have a limited number of possibilities, such as `x = x operator expr`. Only the load and store of the object designated by `x` are atomic; the evaluation of `expr` is not atomic (and should not include any reference to `x`).

Some OpenMP compilers just replace atomic regions by critical regions, usually excluding from the critical region the evaluation of `expr`. Others make use of the efficient machine instructions available in current microprocessors to atomically access memory or through the use of load-linked store-conditional. Figure 2 shows that for a simple example: the compiler generates code so that expression on the right is computed first (in this case, a constant value 1). Once evaluated it "saves" the current version of a(i) and applies the operator (+ in this case). Then

```
!$omp parallel
   do i = 1, n
!$omp atomic
      a(i) = a(i) + 1
   enddo
!$omp end parallel
```

(a)

```
atomic_expr_a = 1 atomic_old_a = a(i) atomic_new_a=atomic_old_a+
atomic_expr_a DO WHILE (0 .EQ.
atomic_update_4(a(i),atomic_old_a,atomic_new_a))
   atomic_old_a = a(i)
   atomic_new_a=atomic_old_a+atomic_expr_a
END DO
```

(b)

Fig. 2. Atomic region in OpenMP and a possible translation

it "tries" to update the shared variable, and if it fails, restores the original value of a(i) and starts again applying the operator. The while loop finishes when the atomic_update_4 function returns successful (different than 0). So notice that this is a simplified version, just for a single variable, of a transaction.

3.2   Is OpenMP CRITICAL a Transaction?

The critical construct in OpenMP restricts execution of the associated structured block to a single thread at a time (among all the threads in the program, without regard to the team(s) to which the threads belong).

```
#pragma omp critical [(name)]
   structured-block
```

An optional name may be used to identify the critical construct. All critical constructs without a name are considered to have the same unspecified name. A thread waits at the beginning of a critical region until no other thread is executing a critical region with the same name. The critical construct enforces exclusive access with respect to all critical constructs with the same name in all threads, not just in the current team.

OpenMP compilers usually replace critical regions with set_lock/unset_lock primitives, declaring a different lock variable for each name used in critical constructs. From this usual implementation and the description above (taken from the current 2.5 language specification) we understand that the code in the structured block can be never speculatively executed by a thread so that several threads are executing it simultaneously.

In this case, a first proposal to integrate TM in OpenMP would be to relax the previous description so that "waiting at the begining of the `critical` region" does not preclude the thread from executing the structured block speculatively as a full transaction. However, we prefered to first propose a set of extensions (construct and clauses) to understant and explore how the TM abstraction could fit with the OpenMP programming style and execution model. We could later see the minimum changes required in the current specification to integrate transactional execution in OpenMP.

3.3 Proposed OpenMP Extensions for TM

The first extension is a pragma to delimit the sequence of instructions that compose a transaction:

```
#pragma omp transaction [exclude(list)|only(list)]
   structured-block
```

With this extension, the programmer should be able to write standard OpenMP programs, but instead of using intrinsic routines to lock/unlock, atomic or critical pragmas, he/she could use this pragma to specify the sequence of statements that need to be executed as a transaction. The optional `exclude` clause can be used to specify the list of variables for which it is not necessary to check for conflicts. This means that the STM library does not need to keep track of them in the read and write sets. On the contrary, if the programmer uses the optional `only` clause, he/she is explicitly specifying the list of variables that need to be tracked. In any case, data versioning for all speculative writes is needed for all shared and private variables in case the transaction needs to be rolled-back.

Another possibility is the use of a new clause associated to the OpenMP worksharing constructs:

```
#pragma omp for transaction [exclude(list)|only(list)]
   for (...;...;...)
      structured-block
```

or

```
#pragma omp sections transaction [exclude(list)|only(list)]
#pragma omp section
   structured-block
#pragma omp section
   structured-block
```

In the first case, each iteration of the loop constitutes a transaction, while in the second case, each section is a transaction.

For OpenMP 3.0, a new tasking execution model is being proposed. Tasks are defined as deferrable units of work that can be executed by any thread in the thread team associated to the active parallel region. Task can create new tasks and can also be nested inside worksharing constructs. In this scenario, data access ordering and synchronization based on locks will be even more difficult to express, so transactions appear as an easy way to express intent and leave the mechanisms to the TM implementation. For tasks we propose the possibility of tagging a task as a transaction, using the same clause specified above.

```
#pragma omp task transaction [exclude(list)|only(list)]
    structured-block
```

3.4   Nebelung Library Interface and Behavior

In order to have a complete execution environment supporting transactional memory, we have implemented our own STM library, named Nebelung. A detailed explanation of the library and its implementation is out of the scope for this paper; therefore we present the relevant issues here. The library satisfies the interface presented in the Figure 3. Nebelung library is typeless (work on a byte level) so we also developed wrapper functions read and write around readtx and writetx, which cast results into the proper types. Note that this interface is similar to the ones provided by other current STM libraries [3].

The library functions have the following semantics: createtx and destroytx create and destroy the required data structures for the execution of a transaction, starttx starts the transaction, committx publishes (i.e., makes visible) the results (writes) of the transaction, aborttx cancels the transaction and retrytx cancels the transaction and restarts it. readtx and writetx are function library function calls for handling memory accesses. The transaction should be started and ended with the code presented in the Figure 4.

The most important parts of the library are surely function calls readtx and writetx and they require special attention. Both functions operate on the byte level. writetx receives the real address where the data should be stored, a size of data and the data itself. Function readtx receives the real address which should be read and the size of the data, and returns the pointer to the

```
Transaction*    createtx    ();
void            starttx     (Transaction *tr);
status          committx    (Transaction *tr);
void            destroytx   (Transaction *tr);
void            aborttx     (Transaction *tr);
void            retrytx     (Transaction *tr);
void*           readtx      (Transaction *tr, void *addr, int blockSize);
void*           writetx     (Transaction *tr, void *addr, void *obj,
                                              int blockSize);
```

Fig. 3. Nebelung library interface to support the code generated by the Mercurium compiler

```
{   Transaction* t = createtx(); while (1) {
    starttx (t);
    if (setjmp (t->context) == TRANSACTION_STARTED) {
```

(a)

```
        if (COMMIT_SUCCESS == committx (t)) break;
        else aborttx (t);
    } else aborttx (t);
  }
  destroytx (t);
}
```

(b)

```
startTransaction();
// transaction body
endTransaction();
```

(c)

Fig. 4. Macros for (a) starting and (b) ending a transaction. (c) Code of the transaction surrounded by the previous macros.

location which holds the requested data. Returned pointer does not need to be the same as the original one and this is implementation dependent. This can be the source of a memory leakage, because it is not clear who should release the pointed memory. There are two possible implementations of function readtx. The first implementation option can copy the data to a new location and return the pointer, requiring the programmer to free it when it is not needed any more. The other option, which we implemented, is that the library takes care about everything and returns the pointer to the location which should be just read. This pointer should not be used later for writing. If the same data should be modified later, that should be done through function writetx and not directly through returned pointer.

The current implementation of Nebelung library performs lazy conflict detection. Read and write sets are maintained dynamically and all memory operations are performed locally for the transaction. At commit time, the library checks if there is any conflict with other transactions. If conflict exists, the current transaction is committed and other transactions are aborted. In this way the transaction progress is guaranteed.

3.5   Source-to-Source Translation in Mercurium

The Mercurium OpenMP source-to-source translator transforms the code inside the transaction block in such a way that for each memory access, a proper

```
#pragma omp transaction {
    ux = (a2->dx -a1->dx)*lambda + (a2->x -a1->x);
    r =one/( ux*ux + uy*uy + uz*uz);
    r0 = sqrt(r);
    ux = ux*r0;
    k = -dielectric*a1->q*a2->q*r;
    r = r*r*r;
    k = k + a1->a*a2->a*r*r0*six;
    k = k - a1->b*a2->b*r*r*r0*twelve;
    a1fx = a1fx + ux*k;
    a2->fx = a2->fx - ux*k;
}
```

(a)

```
{ startTransaction(); {write(t, &ux, (*read(t, &((*read(t,
&a2))->dx))-
    *read(t, &( ( *read(t, &a1))->dx))) *
    *read(t,&lambda)+(*read(t,&((*read(t,&a2))->x))-
    *read(t, &( ( *read(t, &a1)) ->x))));
 write(t, &r,  *read(t, &one) /( *read(t, &ux) *
    *read(t, &ux) + *read(t, &uy) * *read(t, &uy) +
    *read(t, &uz) * *read(t, &uz) ));
 write(t, &r0, sqrt( *read(t, &r) ));
 write(t, &ux,  *read(t, &ux) * *read(t, &r0) );
 write(t, &k, - *read(t, &dielectric) *
    *read(t, &( ( *read(t, &a1)  ) ->q) ) *
    *read(t, &( ( *read(t, &a2)  ) ->q) ) *
    *read(t, &r) );
 write(t, &r,  *read(t, &r) * *read(t, &r) *
    *read(t, &r) );
 write(t, &k,  *read(t, &k) +
    *read(t, &( ( *read(t, &a1))->a) ) *
    *read(t, &( ( *read(t, &a2))->a) ) *
    *read(t, &r)* *read(t, &r0) * *read(t, &six));
 write(t, &k,  *read(t, &k)
    *read(t, &((*read(t, &a1))->b) ) *
    *read(t, &((*read(t, &a2))->b) ) *
    *read(t, &r) * *read(t, &r) * *read(t, &r0) *
    *read(t, &twelve));
 write(t, &a1fx,  *read(t, &a1fx)+ *read(t, &ux)* *read(t, &k) );
 write(t, &( ( *read(t, &a2)  ) ->fx ) ,
    *read(t, &( ( *read(t, &a2)  ) ->fx) )
    *read(t, &ux) * *read(t, &k) );
} endTransaction(); }
```

(b)

Fig. 5. Code generated code for the first critical section in Figure 1 (excerpt from AMMP SpecOMP2001)

```
int f(int); int correct(int* a, int* b, int* x){
    int fx;
#pragma omp transaction exclude (fx) {
    fx = f(*x);
    a += fx;
    b -= fx;
    }
}
```

(original code)

```
int correct(int* a, int* b, int* x){
    int fx;
    { startTransaction();
        {
        fx = f(*read(t, x));
        write(t, &a, *read(t, &a) + fx);
        write(t, &b, *read(t, &b) - fx);
        }
    endTransaction();
    }
}
```

(transactional code)

Fig. 6. Example using the exclude clause

STM library function call is invoked. The current version of Mercurium accepts OpenMP 2.5 for Fortran90 and C.

Figure 5 shows how the first critical region in Figure 1 is specified using our proposed extensions and the code generated by Mercurium. Figure 6 shows a synthetic example using the `exclude` clause.

Finally, Figure 7 shows another example which operates on a binary tree, inserting n nodes into the tree. We are using the current tasking proposal for OpenMP 3.0. Notice that n tasks will be created and executed atomically in parallel. Figure 8 shows the code generated by Mercurium.

3.6 Support for Hardware Transactional Memory

Researchers have not yet built a conscesnus about the best degree of harware support for transactional memory (i.e. full HTM, HaSTM or HyTM), particularly when looking at future multicore architectures with large numbers of cores on a chip and not necessarely cache coherent. This is part of our current work and orthogonal to the transformation process discussed in this paper, which can easily be retargeted to support HTM: we only need to change the `startTransaction` and the `endTransaction` macros. Those macros should use the proper hardware

```
ivoid ParallelInsert(struct BTNode** rootp, int n, int keys[], int
values[]){ #pragma omp parallel single {
   for (int i = 0; i < n; ++i) {
      int key = keys[i], value=values[i];
#pragma omp task capturevalue(key, value) captureaddress(rootp) {
      int inserted, f;
      BTNode* curr;
      BTNode* n;

      n = NewBTNode;
      initNode(n,key,value);
      inserted = 0;
      f = 0;
#pragma omp transaction {
      if (*rootp == 0) {*rootp = n;}
      else {
         curr = *rootp;
         while (inserted == 0) {
            if (curr->key == key){
               curr->value = value;
               curr->valid = 1;
               inserted = 1;
               f = 1;
            } else if (curr->key> key) {
               if (curr->left == 0) {curr->left = n; inserted = 1;}
               else curr = curr->left;
            } else {
               if (curr->right == 0) {
                  curr->right = n;
                  inserted = 1;
               } else curr = curr->right;
            }
         }
      }
      } // end oftransaction
      if (f == 1) free(n);
      } // end of task
   } // end if for loop
   } // end of parallel region
} // end of ParallelInsert function
```

Fig. 7. Function for parallel insertion of n nodes into a binary search tree, expressed using the tasking execution model

ISA instructions for the start and the end of the transaction. The transformation of loads and stores is not needed anymore. To illustrate this, in this paper we chose to incorporate the HTM proposed by McDonald et. al [16] with instructions

```
{ startTransaction(); { if (  *read(t, rootp)  == 0 ) { write(t,
rootp,  *read(t, &n) );
  } else {
     write(t, &curr,  *read(t, rootp));
     while (  *read(t, &inserted)  == 0) {
        if (*read(t, &((*read(t,&curr))->key))== *read(t, &key)) {
           write(t, &((*read(t, &curr))->value), *read(t, &value));
           write(t,&((*read(t,&curr))->valid),1);
           write(t, &inserted, 1);write(t, &f, 1);
        } else if (*read(t,&((*read(t, &curr))->key)) > *read(t, &key)) {
           if(*read(t,&((*read(t, &curr))->left)) == 0 ) {
              write(t, &((*read(t, &curr))->left), *read(t, &n) );
              write(t, &inserted, 1);
           } else
              write(t, &curr, *read(t,&((*read(t,&curr))->left)) );
        } else { if(*read(t,&((*read(t,&curr))->right)) == 0 ) {
              write(t,&((*read(t, &curr))->right ), *read(t, &n) );
              write(t, &inserted, 1);
           } else write(t, &curr, *read(t, &((*read(t,&curr))->right)));
        }
     }
  }
} endTransaction();}
```

Fig. 8. Code generated for the transaction inside the parallelInsert function

```
#define startTransaction() \           #define endTransaction()        \
   { asm { xbegin }                        asm { xvalidate             \
                                                 xcommit              \
                                           }                          \
                                        }
```

Fig. 9. Support for HTM. Definition in pseudo code, of **startTransaction** and **endTransaction** macros in case hardware has instructions **xbegin** and **xcommit** for the start and the end of the transaction, and **xvalidate** for the validation of the transaction.

xbegin, xcommit and xvalidate (denoting the start, end and validation of the transaction, respectively). In this case, the start and end macros would be as shown in Figure 9. Note that **endTransaction** semantics require **xvalidate** since [16] uses two-phase commit.

For those variables that do not need to be tracked, the HTM proposed in [16] includes instructions **imld** and **imst** for load and store without changing read and write sets. Figure 10 shows how the compiler would generate code for a simple a = b + c statement, where c does not need to be tracked, for both STM and HTM.

4   Open Issues

In addition to the HTM implementation challenge mentioned in the previous section, there are other issues that become important challenges when TM is incorporated into OpenMP. Some issues are research challenges while other are restrictions on the way OpenMP and TM can be combined.

The first issue refers to transaction nesting. Nested transactions can be supported in two different ways: closed nested or open nested. In a closed nested TM system either all the transactions that are in a nested region commits or neither. In comparison, in an open nested TM when an inner transaction commits its effects are made visible for all threads in the system. The use of open nested transactions can unleash more concurrency than closed nested ones. When an open nested transaction commits, its write set is made visible to all other transactions, so other transactions can see the modifications sooner and work with this modified data. However open nested transactions increase the burden of the programmer: compensating actions are needed when the outermost transaction commits and when one of the surrounding transactions aborts. The handling of this compensating code could be quite complex, and the programmer must have an expert level grasp of the semantics of the code. For this reason, although OpenMP-TM can incorporate closed-nested transactions relatively easily, open-nested transactions are challenging because the compensating code could impact the sequential nature of the program when it is compiled without OpenMP.

```
a = b + c;       write(t, &a, *read(&b) + c);     asm {
                                                      load   r0, b
                                                      imld   r1, c
                                                      add    r0, r1
                                                      store  a,  r0
                                                   }
```

(original code)   (transformed STM code)   (transformed HTM code)

Fig. 10. Transformation of a simple statement using ii) STM or iii) HTM in pseudo assembler code

The second issue refers to the use of I/O inside a transaction. I/O inside a critical section is not a problem for OpenMP because such blocks are protected and never rolled back. However, for TM the issue is different: suppose that inside a transaction, a system call attempts to output a character to the terminal. One solution is to execute the system call immediately; however it would be very problematic if this transaction aborts later. Trying to "undo" the I/O by deleting the character upon an abort would obviously lead to a very wobbly system. In some cases, even executing the I/O operation may be difficult if the data is buffered in HTM. A different approach to solve the I/O problem would be to categorize I/O based on their abortive properties. By definition, an I/O call is undoable if its effects could be rolled back which in turn implies that its effects are self-contained to the I/O operation only. Here, the challenge is to allow maximum programmer expressiveness while avoiding too complex implementations. Inevitably, programmers must be aware of certain kinds of I/O operation that simply do not make sense to transparently perform as part of a single atomic transaction: for instance, prompting the user for input and then receiving and acting on the input. However note that those compensating actions can also affect the sequential nature of the program when it is compiled without OpenMP. We believe that a first-order approach is to forbid the I/O operations in OpenMP-TM transactions.

Another issue is about the nesting of OpenMP constructs and transactions. There are two cases that one needs to consider. In the first case, a transaction might exist inside an OpenMP parallel region or worksharing construct. Note that the transaction might be invoked from within a library call, so the programmer might not be aware of this. However, even with this added complexity, this case is easy to handle: each thread in the team will be having a transaction inside. The other case, where an OpenMP parallel or worksharing construct appears inside the scope of a transaction, is more complicated. Note that for this case as well, the programmer may not be aware that an OpenMP construct exists within the transaction; this can happen if the programmer uses library function inplemented in OpenMP. The complexity rises from the fact any of the OpenMP threads are liable to be aborted at any time. This signifies that if one created thread is aborted, then all the other threads must also abort unnecessarily due to the semantics of TM. This can be avoided using a mechanism like close-nested transaction allowing to rollback only the conflicting thread if it is possible to ensure that the conflict only affects this thread.

Finally, it would be interesting to include additional functionalities to the basic transactional execution model offered by **transaction**. For example, we could specify a condition; if the evaluation of the condition returns false then the transaction is known to abort. Similarly, we could add a condition that needs to be evaluated once a transaction is aborted; if the evaluation of the condition returns false, it is known that a reexecution of the transaction will fail again. So it is better no to execute it until the condition is true. These conditions could for example include operators to check that a shared variable has been accessed (touch(var_name)).

```
#pragma omp transaction [retrywhen(condition)] [executeif(condition)]
    structured-block
```

5   Conclusions

In this paper we have done a first exploration of how Transactional Memory (TM) could provide a more graceful and natural mechanism to write parallel programs than using lock-based mechanisms. Recently there is a flurry of research activity to define and design Hardware and Software Transactional Memory, and our OpenMP community should follow this activity in order to take benefit as soon as possible. This paper has covered a basic proposal to extend OpenMP with transactions and identified some significant future challenges in the OpenMP-TM tandem. However, the biggest challenge is to make the adoption of Transactional Memory by the (OpenMP) programmer community as smooth as possible through HW/SW design; effective mechanisms for supporting TM are crucial to fulfilling the promise of improved application performance on future many-core CMPs.

During the preparation of this version of the paper, we realized that a paper on a similar topic was accepted for publication [17]. However, only the abstract of the other paper was available to us, so we were not able to make any comparisons at this point.

Acknowledgments. This work is supported by the cooperation agreement between the Barcelona Supercomputing Center National Supercomputer Facility and Microsoft Research, by the Ministry of Science and Technology of Spain and the European Union (FEDER funds) under contract TIN2004-07739-C02-01 and by the European Network of Excellence on High-Performance Embedded Architecture and Compilation (HiPEAC).

References

1. Larus, J., Rajwar, R.: Transactional Memory. Morgan Claypool, San Francisco (2006)
2. OpenMP Architecture Review Board, OpenMP Application Program Interface (May 2005)
3. Harris, T., Plesko, M., Shinnar, A., Tarditi, D.: Optimizing Memory Transactions. In: PLDI 2006. ACM SIGPLAN 2006 Conference on Programming Language Design and Implementation (June 2006)
4. Allen, E., Chase, D., Luchangco, V., Maessen, J.-W., Ryu, S., Steele Jr., G.L., Tobin-Hochstadt, S.: The Fortress Language Specification. Sun Microsystems (2005)
5. Charles, P., Grothoff, C., Saraswat, V., Donawa, C., Kielstra, A., Ebcioglu, K., von Praun, C., Sarkar, V.: X10: an Object-oriented approach to non-uniform Cluster Computing. In: Proceedings of the 20th Annual ACM SIGPLAN Conference on Object-oriented Programming Systems Languages and Applications (OOPSLA), New York, USA, pp. 519–538 (2005)
6. Cray. Chapel Specification (February 2005)
7. Shavit, N., Touitou, D.: Software Transactional Memory. In: Proceedings of the 14th Annual ACM Symposium on Principles of Distributed Computing, pp. 204–213 (1995)
8. Herlihy, M., Eliot, J., Moss, B.: Transactional Memory: Architectural Support for Lock-Free Data Structures. In: Proc. of the 20th Int'l Symp. on Computer Architecture (ISCA 1993), May 1993, pp. 289–300 (1993)
9. Damron, P., Fedorova, A., Lev, Y., Luchangco, V., Moir, M., Nussbaum, D.: Hybrid Transactional Memory. In: Proceedings of the Twelfth International Conference on Architectural Support for Programming Languages and Operating Systems (ASPLOS) (October 2006)

10. Kumar, S., Chu, M., Hughes, C.J., Kundu, P., Nguyen, A.: Hybrid Transactional Memory. In: Proceedings of ACM Symp. on Principles and Practice of Parallel Programming (March 2006)
11. Saha, B., Adl-Tabatabai, A., Jacobson, Q.: Architectural Support for Software Transactional Memory. In: 39th International Symposium on Microarchitecture (MICRO) (2006)
12. Shriraman, A., Marathe, V.J., Dwarkadas, S., Scott, M.L., Eisenstat, D., Heriot, C., Scherer III, W.N., Spear, M.F.: Hardware Acceleration of Software Transactional Memory. In: TRANSACT 2006 (2006)
13. Balart, J., Duran, A., Gonzàlez, M., Martorell, X., Ayguadé, E., Labarta, J.: Nanos Mercurium: A Research Compiler for OpenMP. In: European Workshop on OpenMP (EWOMP 2004), Stockholm, Sweden, October 2004, pp. 103–109 (2004)
14. Martorell, X., Ayguadé, E., Navarro, N., Corbalan, J., Gonzalez, M., Labarta, J.: Thread Fork/join Techniques for Multi-level Parallelism Exploitation in NUMA Multiprocessors. In: 13th International Conference on Supercomputing (ICS 1999), Rhodes (Greece) (June 1999)
15. Milovanović, M., Unsal, O.S., Cristal, A., Stipić, S., Zyulkyarov, F., Valero, M.: Compile time support for using Transactional Memory in C/C++ applications. In: 11th Annual Workshop on the Interaction between Compilers and Computer Architecture INTERACT-11, Phoenix, Arizona (February 2007)
16. McDonald, A., Chung, J., Carlstrom, B., Minh, C., Chafi, H., Kozyrakis, C., Olukotun, K.: Architectural Semantics for Practical Transactional Memory. In: Proc. 33th Annu. international symposium on Computer Architecture, pp. 53–65 (2006)
17. Baek, W., Minh, C.-C., Trautmann, M., Kozyrakis, C., Olukotun, K.: The OpenTM Transactional Application Programming Interface. In: Proc. 16th International Conference on Parallel Architectures and Compilation Techniques (PACT 2007), Romania (September 2007)

OpenMP on Multicore Architectures

Christian Terboven, Dieter an Mey, and Samuel Sarholz

Center for Computing and Communication,
RWTH Aachen University
{terboven,anmey,sarholz}@rz.rwth-aachen.de

Abstract. Dualcore processors are already ubiquitous, quadcore processors will spread out this year, systems with a larger number of cores exist, and more are planned. Some cores even execute multiple threads. Are these processors just SMP systems on a chip? Is OpenMP ready to be used for these architectures? We take a look at the cache and memory architecture of some popular modern processors using kernel programs and some application codes which have been developed for large shared memory machines beforehand.

1 Introduction

Chip multi-processing (CMP) already is ubiquitous. The first dualcore processor was IBM's Power4 chip, which was introduced in 2001, and Sun's UltraSPARC-IV processor followed in 2004. Since AMD and Intel introduced their first dualcore processor chips in 2005 and 2006 respectively, discounters are selling PCs equipped with dualcore processors. Today, Intel's first quadcore processor (code named Clovertown) is available and AMD will follow later this year. Some cores even execute multiple threads (CMT=chip multi-threading), the first popular implementation was Intel's Hyper-Threading technology. Sun's UltraSPARC T1 processor (code named Niagara) provides 8 cores running 4 threads each, but offers only little floating point capabilities.

Multicore processors provide more compute power for less energy and produce less heat, which has become the delimiting factor for building processors with higher clock rates. As a consequence parallelization is getting even more important. Thus the questions arise, whether these processors are just SMP systems on a chip and whether OpenMP is ready for these architectures.

We take a look at some popular modern processors which we describe in chapter 2. Using some kernel programs, we investigate the cache and memory architecture of these machines in chapter 3 and then take a look at some application codes which have been parallelized with OpenMP for large shared memory machines beforehand in chapter 4 before we conclude in chapter 5.

2 The Machinery

For our performance analysis experiments we used several dualcore and one quadcore machine. It turned out that these machines have different characteristics regarding memory latency, memory bandwidth and OpenMP scheduling

B. Chapman et al. (Eds.): IWOMP 2007, LNCS 4935, pp. 54–64, 2008.

Table 1. Machines (with abbreviations) used for performance analysis

Machine / year of installation	Processor / Frequency [GHz]	Abbrev.	Sockets	Chips per Socket	Cores per Chip	Threads per Core	Level 2 Cache
Noname 2003	Intel Pentium4 Xeon 2.66	XeonHT	2	1	1	2	0.5 MB / chip
Sun Fire E2900 2004	Sun Ultra- SPARC IV 1.2	USIV	12	1	2	1	8 MB / core
Sun Fire T2000 2005	Sun Ultra- SPARC T1 1.0	Niagara	1	1	8	4	3 MB / chip
Sun Fire V40z 2005	AMD Opteron875 2.2	Opteron	4	1	2	1	1 MB / core
Dell Power- Edge 1950 2006	Intel Xeon 5160 3.0	Woodcrest	2	1	2	1	4 MB / chip
Dell Power- Edge 1950 2007	Intel Xeon 5355 2.66	Clovertown	2	2	2	1	4 MB / chip

overhead. Table 1 summarizes the architecture details and introduces abbreviations that are used throughout the rest of this paper.

The machines differ in how the memory is organized. Whereas the USIV, Niagara, Woodcrest and Clovertown machines have a rather flat memory system, the Opteron-based machine has a ccNUMA architecture. Whereas the local memory access is very fast on such a system, multiple simultaneous remote accesses can easily lead to grave congestions of the HyperTransport links.

The processors investigated here differ in how their level 2 (L2) cache is organized. The USIV and the Opteron processors have one cache per core, whereas the Niagara and the Woodcrest processors both have an L2 cache that is shared by all cores. The Clovertown integrates two Woodcrest chips onto one socket and therefore it has two L2 caches which are shared by two cores each. Table 1 also lists the L2 cache sizes and their association.

On the Intel-based machines the Linux operating system is running, on the SPARC- and Opteron-based machines the Solaris operating system is running. On Linux, we used the Intel 10.0 beta (release 0.013) compilers, as they introduced explicit optimizations for the Woodcrest and Clovertown processors. On Solaris, we used the Sun Studio Express (build 35_2) compilers, again these pre-production compilers delivered the best performance.

The operating systems present the multicore processors to the user as if there were multiple singlecore processors in the system. However, they differ in how the cores are numbered. For a dual-socket dualcore processor system, the Solaris operating system assigns the virtual processor ids 0 and 1 to the first socket and

2 and 3 to the second socket. The Linux operating system on the other hand, assigns 0 and 2 to the first socket and 1 and 3 to the second socket. On the quadcore Clovertown system, the virtual processor ids 0 to 3 are assigned to the first socket and 4 to 7 are assigned to the second socket. However, the ids 0 and 2 share one cache, for example.

3   Memory Performance

3.1   Memory Latency and Bandwidth

Latency. The memory latency can be measured using a technique named *pointer chasing* by a long sequence of load instructions, as implemented by the following code snippet:

```
p = (char **)*p;
```

The content of the memory location just read from memory delivers the address of the following load instruction. Thus, these load instructions cannot be overlapped. The addresses have a stride of -64 causing the load of an L2 cache line for each load instruction.

Bandwidth. The next code snippet shows the operation that is timed in order to measure the memory bandwidth:

```
long long *x, *xstart, *xend, mask;
for ( x = xstart; x < xend; x++ ) *x ^= mask;
```

So each loop iteration involves one load and one store of a variable of type long long.

We ran an OpenMP-parallelized version of both kernels in order to find out in how far multiple threads interfere with each other and we used explicit processor binding to carefully placed the threads onto the processor cores. The threads work on large ranges of private memory which do not fit in the L2 caches, so each load and store operation goes to main memory (see table 2).

Table 2 contains a column "Boards-Sockets-Chips-Cores-Threads" which describes how we distributed the OpenMP threads onto the machines. For example "1-2-1-2-1" signifies that we ran 4 OpenMP threads using 1 board, 2 sockets on this board, 1 chip on each socket (Clovertown has 2 chips per socket), 2 cores per chip and 1 thread per core.

The table indicates that the AMD and Intel processors profit from hardware prefetches when accessing successive memory chunks of 64 bytes in reverse order. Interestingly, the memory latency does not suffer much when increasing the stress on the memory systems of the machines. Four threads on the Woodcrest experience a quadruplication and 8 threads on the Clovertown experience an 8-fold increase of the memory latency.

Both the Woodcrest and the Clovertown machines display a high memory bandwidth for one thread. When running two threads the achievable memory bandwidth already heavily depends on their placement. On the Woodcrest it is profitable to run two threads on two chips which each have their own socket. On the Clovertown it is profitable to run two threads on both chips residing on the same socket, whereas if two threads run on separate sockets, the bandwidth drops by a factor of 1.5. Two threads running on both cores of a single chip clearly experience a memory bandwidth bottleneck on both machines. On the other hand two threads are able to almost consume the maximum available memory bandwidth on these machines.

The USIV, Niagara and Opteron machines provide less memory bandwidth for one thread, but display a better scalability.

The placement of the data is an important issue on the Opteron machine, which has a pronounced ccNUMA architecture. These tests clearly profit from a careful usage of the first touch policy compared to a quasi-random placement.

The memory bandwidth of the Niagara machine seems to be optimal with 16 threads running on 4 out of 8 cores - a fact which deserves further investigations.

The older XeonHT machine clearly suffers from a known memory bottleneck which prohibits any scalability.

3.2   Matrix Transposition

Matrix transposition involves no floating point operations, but stresses the memory bandwidth. A square matrix with 8 byte floating point numbers large enough to not fit in the available caches is read once and written once. When programming in C, matrix rows are stored consecutively in memory. Reading and writing rows leads to nice memory access patterns using whole cache lines efficiently, whereas accessing the matrix column wise, only 8 bytes are used out of each cache line, which are 64 byte for all machines under investigation. Obviously, the algorithm which we are using can be improved by blocking. But we want to find out, how the processors' memory hierarchies cope with unfavorable accesses. In many cases we experience quite some variations in the measurements depending on the memory footprint. This kind of experiments is quite sensitive to peculiarities of the cache architecture, like cache associativity effects. Here we do not want to focus on such details but just look at the memory performance in relation to the thread placement and the loop scheduling. Therefore the table contains memory bandwidth ranges of the matrix transpose as have been measured in our experiments.

USIV. With one thread the obtainable memory bandwidth is about 0.7 GB/s. With more threads the memory performance depends on the loop schedule. A static schedule for 2 threads leads to a slight improvement (approx. 0.95 GB/s) because the matrix transpose leads to a load imbalance, such that the slave thread has some idle time. This load imbalance can be easily cured by a suitable loop schedule like guided,8, whereas the dynamic schedule adds too much overhead. The USIV machine reveals a very flat memory system and thread placement only has a minor effect.

Niagara. A single thread only obtains a bandwidth of about 0.4 GB/s. When running up to 8 threads on different cores, the bandwidth nicely scales, whereas

Table 2. Latency and Bandwidth measurements of multicore architectures

Architecture	#Threads	Boards-Sockets-Chips-Cores-Threads	Latency [nsec]	Accum. Bandwidth [MB/s]	Matrix Transpose Best effort Schedule	Bandwidth
USIV	1	1-1-1-1-1	233	1609	static	0.7
	2	1-1-1-2-1	249	3100	guided 1/8	1.4
	2	1-2-1-1-1	233	3137	guided 1/8	1.4
	4	1-2-1-2-1	250	4333	guided 1/8	2.3
	4	1-4-1-1-1	234	5295	guided 1/8	2.4-2.5
	8	1-4-1-2-1	251	6402	guided 1/8	3.5-4.0
	8	2-4-1-1-1	240	6056	guided 1/8	3.4-3.6
	16	2-4-1-2-1	260	7006	guided 1/8	4.4-5.0
Niagara	1	1-1-1-1-1	108	845	static	0.38-0.42
	2	1-1-1-1-2	110	1580	dynamic,1	0.41-0.46
	2	1-1-1-2-1	110	1660	guided,1/8	0.76-0.85
	4	1-1-1-1-4	110	2770	dynamic,1	0.45-0.49
	4	1-1-1-2-2	110	3156	dynamic,1	0.84-0.94
	4	1-1-1-4-1	110	3451	guided 1/8	1.52-1.70
	8	1-1-1-2-4	111	5464	dynamic,1	0.9-1.0
	8	1-1-1-4-2	111	6216	dynamic,1	1.76-1.87
	8	1-1-1-8-1	111	6512	dynamic,1	3.36-3.38
	16	1-1-1-4-4	115	10560	dynamic,1	1.86-1.97
	16	1-1-1-8-2	115	6992	dynamic,1	3.51-3.69
	32	1-1-1-8-4	131	4672	dynamic,1	3.58-3.77
Opteron	1	1-1-1-1-1	34	3497	static	2.2-2.6
	2	1-1-1-2-1	36	4629	guided 1/8	3.1
	2	1-2-1-1-1	36	6860	guided 1/8	3.1
	4	1-2-1-2-1	36	9223	static / guided	2.9
	4	1-4-1-1-1	36	13448	static / guided	2.9-3.2
	8	1-4-1-2-1	40	18307	guided 1/8	2.7-2.9
Woodcrest	1	1-1-1-1-1	22	4583	static	2.6
	2	1-1-1-2-1	32	4727	guided 1/8	4.0
linuxwcc00	2	1-2-1-1-1	24	7903	guided 1/8	4.8
	4	1-2-1-2-1	42	7997	guided 1/8	6.50-7.00
Clovertown	1	1-1-1-1-1	24	3970	static	2.3
	2	1-1-1-2-1	36	3998	guided 1/8	3.4
	2	1-1-2-1-1	28	6871	guided 1/8	4.1
	2	1-2-1-1-1	32	4661	guided 1/8	3.8
	4	1-1-2-2-1	66	6887	guided 1/8	6.0
	4	1-2-2-1-1	44	8009	guided 1/8	6.7
	4	1-2-1-2-1	62	4660	guided 1/8	4.1
	8	1-2-2-2-1	86	8006	guided 1/8	6.7
XeonHT	1	1-1-1-1-1	28	2559	static	1.3
	2	1-1-1-1-2	42	1956	dyn/guided	1.10-1.30
	2	1-2-1-1-1	44	1930	guided 1/8	1.10-1.30
	4	1-2-1-1-2	90	1710	static	1.20

the bandwidth does not increase much when using multiple threads per core. The `static` schedule clearly suffers from the load imbalance whereas overhead of the `dynamic` schedule is not visible. False sharing effects are not observable due to the sharing of the on-chip L2 cache. In fact, the schedule `dynamic,1` slightly outperforms the others in most cases.

XeonHT. The Pentium 4 Xeon architecture has a known memory bottleneck, which restricts any speed-up of memory hungry parallel applications. Also the memory transpose kernel does not profit from parallelization. When comparing the performance of two threads running on one processor with Hyper-Threading to two threads on two processors, it can be seen again, that sharing the caches reduces the negative impact of false sharing. All in all, parallelization does not pay off here.

Opteron. As we did not focus on ccNUMA effects of the quad-socket dualcore Opteron machine here (see [1]) we let the whole matrix be allocated close to the master thread by the default first-touch policy of the Solaris operating system. Thus, the master thread obtains a memory bandwidth of about 2.6 GB/s. Two threads on a single chip or on two different chips hardly experience any difference, as both cores do not share any caches. Using a `guided` schedule the bandwidth increases to 3.1 GB/s. But when using 4 or more threads the bandwidth slightly decreases, most likely because of the saturation of the HyperTransport links. As there is no sharing of caches, the machine is sensitive to false sharing effects, no matter where the threads are running. Also small chunksizes hurt and the overhead of the `dynamic` schedule is clearly visible. We employed large pages, as supported by the Solaris operating system, to reduce the number of TLB misses.

Woodcrest. Both dualcore Woodcrest processors access the main memory via a common front side bus. Whereas a single thread obtains about 2.6 GB/s bandwidth and two threads distributed over the two sockets up to 5 GB/s, 4 threads are limited to about 6.5-7 GB/s in this test case. Two threads running on the same processor obtain some 4 GB/s.

Two threads on both processors profit from an increased bandwidth and two caches, but they suffer from false sharing effects, when an unfavorable loop schedule like `dynamic,1` is employed, whereas false sharing effects are not observable, if both threads have access to the same L2 cache.

The `guided` schedule delivers the best results for large matrices in all cases. Two threads running on one processor chip perform extremely well when the matrix fits into the shared L2 cache. If the matrix has a memory footprint of 1 MB, 2 threads on one processor (best effort: `guided` schedule) outperform 4 threads on 2 processors (best effort: `static` schedule) by more than a factor of 3. Even a single thread is about 2 times faster than 4 threads in such a case.

Clovertown. For our investigations the Clovertown is the most interesting architecture, as it allows to compare effects of threads sharing a cache and threads with separate caches on the same socket. The Clovertown processor reveals similar attitudes as the Woodcrest processor in that each pair of two cores share the same L2 cache, such that threads sharing one cache, don't suffer from false sharing and perform extremely well if the matrix fits into that cache. Other than that, two threads on cores of one socket which do not share the L2 cache perform very similar to two threads on cores of two different sockets. The `guided` schedule is preferable for large matrices and a small chunksize leads to false sharing if more than one L2 cache is involved.

4 Application Performance

4.1 Sparse Matrix-Vector-Multiplication

A sparse matrix vector product is the most time consuming part in CG-type linear equation solvers, as used in many PDE solvers. Compared to dense linear algebra functions, the data structure used to exploit the matrix sparsity pattern typically adds one level of indirection to the operation.

Here, we used the sparse matrix vector multiplication kernel and the matrices of the C++ Navier-Stokes solver DROPS [1]. A lot of effort has been invested to tune and parallelize this kernel on several different architectures. We compare two matrix sizes: A large matrix with 19,593,418 nonzeros in 23,334 rows taking about 150 MB of memory and a small matrix with 784,075 nonzeros in 11,691 rows taking about 6 MB of memory. Table 3 shows selected performance measurements of this benchmark.

Table 3. Sparse Matrix Vector Multiplication

Abbrev.	#Threads	Sockets-Chips-Cores-Threads	MFLOP/s large / small	Abbrev.	#Threads	Sockets-Chips-Cores-Threads	MFLOP/s large / small
USIV	1	1-1-1-1	141.0 / 203.6	Woodcrest	1	1-1-1-1	557.4 / 633.2
	2	1-1-2-1	274.8 / 426.4		2	2-1-1-1	872.9 / 1442
	2	2-1-1-1	243.7 / 418.8		2	1-1-2-1	643.5 / 776.9
	4	2-1-2-1	423.8 / 821.6		4	2-1-2-1	951.3 / 2147
Opteron	1	1-1-1-1	371.8 / 383.0	XeonHT	1	1-1-1-1	400.0 / 429.4
	2	1-1-2-1	643.5 / 669.1		2	1-1-1-2	442.5 / 445.1
	2	2-1-1-1	721.6 / 743.9		2	2-1-1-1	413.0 / 391.7
	4	2-1-2-1	1200 / 1254		4	2-1-1-2	293.1 / 279.4
Clovertown	1	1-1-1-1	495.2 / 564.6	Niagara	1	1-1-1-1	40.0 / 41.2
	2	1-2-1-1	764.3 / 1420		2	1-1-1-2	72.3 / 75.3
	2	2-1-1-1	622.5 / 1338		2	1-1-2-1	77.6 / 81.0
	2	1-1-2-1	563.9 / 641.3		4	1-1-1-4	118.1 / 122.2
	4	2-2-1-1	902.9 / 4798		4	1-1-4-1	149.9 / 157.5
	4	2-1-2-1	632.6 / 1112		8	1-1-8-1	292.2 / 309.0
	4	1-2-2-1	807.9 / 1943		16	1-1-8-2	520.2 / 544.8
	8	2-2-2-1	913.2 / 9340		32	1-1-8-4	739.8 / 678.1

- USIV: There is no visible performance difference in using two threads on two different sockets or on one socket with two cores, as each core has its own cache. One can notice a super-linear speedup with two threads for the small dataset as the matrix and the vectors then fit into the two L2 caches.

- Opteron: Using two threads on two different sockets is faster than on two cores on one socket, as the threads already can saturate the local memory bandwidth of one socket.
- Woodcrest: Because two cores in one chip on one socket share the L2 cache, thread placement has a strong impact on performance. By using two cores on one socket for two threads, the performance increase is quite small compared to using just one thread. If two threads on two different sockets can utilize one full L2 cache each, super-linear speedup can be observed.
- Clovertown: The variation in performance depending on the thread placement is drastic. Placing two threads on the two cores of a chip results in a speedup of only 1.14 while placing them onto two chips on the same socket results in a speedup of 1.54, for the large dataset. The Clovertown really mimics the behavior of a two socket Woodcrest system. If the matrix fits into the caches, one can see a super-linear speedup of 16 using 8 threads.
- XeonHT: Hyper-Threading is slightly profitable, as computation and waiting for data to arrive from memory can be overlapped, but it is far less efficient than CMT as implemented in the Niagara processor. The path to the memory, which is shared by both sockets, prevents scalability.
- Niagara: On the Niagara processor, we ran a modified version of the benchmark, replacing the 64bit double data type by 64bit integer. On this architecture, CMT proves to be profitable. As the L2 cache is shared by all cores, the performance for the small and the large dataset is about the same. Using 32 threads the large dataset shows a higher scalability than the small one, as the ratio of time waiting for the memory to time spent in computation increases.

In a multi-user environment an application may run on an unsymmetric part of a machine, which may heavily impact performance. We tried to simulate such a situation by using 5 threads on a Clovertown processor. Then there are two threads running on one chip and sharing the available cache and memory bandwidth. It turned out, that in such a situation the performance can be improved from 2216.5 MFLOP/s to 4276.3 MFLOP/s when replacing the original `static,128` schedule by a `dynamic` schedule!

As the Niagara processor has just one cache, the architecture is much more symmetric and the difference between sharing one core by multiple threads and using one core exclusively is far less significant.

4.2 Contact Analysis of Bevel Gears

In the Laboratory for Machine Tools and Production Engineering of the RWTH Aachen University, the contact of bevel gears is simulated and analyzed in order to e.g. understand the deterioration of differential gears as they are used in car gearboxes. These simulations usually run for a few days when using the original serial code. The program was parallelized using OpenMP and it turned out that it scales quite well on multicore architectures, as it is very cache friendly. The parallel code versions consists of some 90,000 lines of Fortran90 code containing 5 parallel OpenMP regions and 70 OpenMP directives.

Table 4. Runtime and Speedup of bevel gear contact analysis

Sockets-Chips-	USIV		Opteron		Woodcrest		Clovertown		XeonHT	
Cores-Threads	Time	Speedup	Time	Speedup	Time	Speedup	Time	Speedup	Time	Speedup
1-1-1-1	276.33	1.0	57.96	1.0	36.02	1.0	48.1	1.0	180.55	1.0
1-1-1-2	-	-	-	-	-	-	-	-	175.43	1.03
1-1-2-1	161.62	1.71	34.91	1.66	21.71	1.66	28.1	1.71	-	-
1-2-1-1	-	-	-	-	-	-	28.1	1.71	-	-
2-1-1-1	162.95	1.70	34.79	1.67	21.73	1.66	28.0	1.72	116.75	1.55
2-1-1-2	-	-	-	-	-	-	-	-	116.00	1.56
1-2-2-1	-	-	-	-	-	-	16.1	2.99	-	-
2-1-2-1	90.87	3.04	20.25	2.86	14.64	2.46	16.1	2.99	-	-
2-2-1-1	-	-	-	-	-	-	16.47	3.01	-	-
4-1-1-1	91.25	3.03	20.11	2.88	-	-	-	-	-	-
2-2-2-1	-	-	-	-	-	-	10.0	4.81	-	-
4-1-2-1	51.86	5.33	12.43	4.66	-	-	-	-	-	-

Table 4 shows that the application scales rather well on all machines. Only Hyper-Threading on the Xeon machine does not pay off. For such a cache friendly application the multicore architectures work very well and thread placement effects are neglectable.

4.3 Simulation of Deformable Objects in VR

In order to meat the real time requirement of a physically based simulation in the context of an interactive virtual environment, in [2] three versions of an FEM solver were parallelized with OpenMP and the scalability on the Opteron, Woodcrest and Clovertown processors have been analyzed. It turned out that this application profits from the L2 caches which are shared by two cores each. On the Opteron synchronization between threads has to involve the main memory, and therefore the speedup is not as good as the speedup of the Clovertown (see figure 1). The best achievable speedup was 6.5 with 8 threads on the Clovertown system.

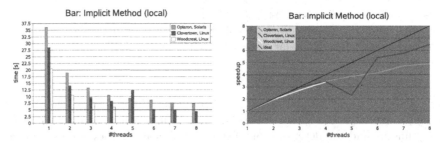

(a) Runtime in seconds for 5 seconds simulation time (red line).

(b) Corotational FEM: Speedup.

Fig. 1. Speedup of Deformable Object Simulation

4.4 ThermoFlow

The Finite Element CFD solver has been developed at the Jet Propulsion Laboratory at the RWTH Aachen University to simulate heat transfer in rocket combustion chambers. The current OpenMP code version has 29,000 lines of Fortran and 69 parallel loops with the whole compute intense part contained in one parallel region. Addressing long vectors continuously as well as indirectly heavily stresses the memory system. Whereas the large Sun Fire SMP systems, which have been the target architecture for this code version, have a scalable memory system (see [3]), the new Intel and AMD multicore processors show only a limited speedup.

Table 5. Runtime and Speedup of ThermoFlow and TFS

Architecture	#Threads	Sockets-Chips-Cores-Threads	ThermoFlow Runtime [sec]	Speedup	TFS Runtime [sec]	Speedup
USIV	1	1-1-1-1	66.4	1.0	33.6	1.0
	2	1-1-2-1	42.8	1.55	18.8	1.79
	2	2-1-1-1	42.2	1.57	18.9	1.78
	4	2-1-2-1	28.9	2.30	11.4	2.95
Opteron	1	1-1-1-1	29.2	1.0	14.3	1.0
	2	1-1-2-1	20.7	1.41	11.6	1.23
	2	2-1-1-1	18.5	1.58	11.6	1.23
	4	2-1-2-1	13.6	2.15	8.7	1.64
Woodcrest	1	1-1-1-1	20.1	1.0	10.0	1.0
	2	2-1-1-1	15.0	1.34	6.0	1.67
	2	1-1-2-1	16.1	1.25	8.0	1.25
	4	2-1-2-1	13.0	1.55	5.0	2.0
Clovertown	1	1-1-1-1	23.1	1.0	11.1	1.0
	2	1-2-1-1	17.0	1.36	7.0	1.59
	2	1-1-2-1	19.0	1.22	9.1	1.22
	2	2-1-1-1	17.1	1.35	8.0	1.39
	4	2-2-1-1	13.1	1.76	4.1	2.71
	4	2-1-2-1	15.1	1.53	7.1	1.56
	4	1-2-2-1	14.1	1.64	6.0	1.85
	8	2-2-2-1	13.0	1.78	4.1	2.71

4.5 The Navier Stokes Solver TFS

The Navier Stokes Solver TFS is a toolkit for CFD simulations developed at the Aerodynamic Institute of the RWTH Aachen University (see [4]). Here we run a benchmark version, which has been parallelized on the inner loop level. As the 3-dimensional computational grid is mapped to 1-dimensional arrays, frequent consecutive accesses to such long vector flush the caches and consume a high memory bandwidth. Like in chapter 4.4, the USIV processor shows better speed-up than the other architectures. Here, two threads sharing one common cache of the Intel processors display a decent speedup. As the memory access is the bottleneck, 8 cores of the Clovertown do not perform better than 4 cores.

5 Conclusion

From the perspective of the application programmer, a multicore processor is an SMP on a chip and OpenMP programs just run nicely. Of course there is a lot more sharing of resources compared to previous SMP systems, which may heavily impact performance. The cache architecture plays a major role. Sharing of caches can be a major advantage, as has been shown in chapters 3.2 and 4.3, where false sharing effects might disappear under lucky circumstances. On the other hand, the memory bandwidth bottleneck is getting even worse with more cores demanding more data, thus limiting the speed-up of memory hungry OpenMP application programs (see chapters 4.5 and 4.4).

Experiences with chip multi-threading have been reported at least since Intel introduced Hyper-Threading and in many cases the speedup gain was quite limited. On the Sun UltraSPARC T1 processor we observed that CMT can be implemented more efficiently (see chapter 3).

The optimal choice for loop scheduling may heavily depend on the machine architecture, on the machine load and on the placement of threads onto the cores. This is a strength of OpenMP, as changing the schedule just means setting another keyword into the schedule clause. Thus it might be convenient to be able to change the schedule `runtime` not only by an environment variable, but also by a runtime call. An additional self tuning schedule, which would allow the OpenMP runtime system to dynamically adapt the schedule over time, is desirable.

Future Work. Plans for future work include looking at additional processors: IBM's Power5 and Intel's Itanium2. Furthermore we want to experiment with self adaptive scheduling schemes, which may be profitable in a multi-user environment.

References

1. Terboven, C., Spiegel, A., an Mey, D., Gross, S., Reichelt, V.: Parallelization of the C++ Navier-Stokes Solver DROPS with OpenMP. In: Proc. ParCo 2005, Malaga, Spain, September 2005. NIC book series, Research Centre Jülich, Germany, vol. 33 (2005)
2. Jerabkova, L., Terboven, C., Sarholz, S., Kuhlen, T., Bischof, C.: Exploiting Multicore Architectures for Physically Based Simulation of Deformable Objects in Virtual Environments. In: 4th workshop on Virtuelle und erweiterte Realität, Weimar, Germany (July 2007) (accepted)
3. an Mey, D., Haarmann, T.: Pushing Loop-Level Parallelization to the Limit. In: EWOMP 2002, Rome, Italy (September 2002)
4. Hörschler, I., Meinke, M., Schröder, W.: Numerical simulation of the flow field in a model of the nasal cavity. Computers & Fluids 32, 3945 (2003)

Supporting OpenMP on Cell

Kevin O'Brien, Kathryn O'Brien, Zehra Sura, Tong Chen, and Tao Zhang

IBM T.J. Watson Research Center, Yorktown Heights, NY 10598

Abstract. The Cell processor is a heterogeneous multi-core processor with one Power Processing Engine (PPE) core and eight Synergistic Processing Engine (SPE) cores. Each SPE has a directly accessible small local memory (256K), and it can access the system memory through DMA operations. Cell programming is complicated both by the need to explicitly manage DMA data transfers for SPE computation, as well as the multiple layers of parallelism provided in the architecture, including heterogeneous cores, multiple SPE cores, multithreading, SIMD units, and multiple instruction issue. There is a significant amount of ongoing research in programming models and tools that attempts to make it easy to exploit the computation power of the Cell architecture. In our work, we explore supporting OpenMP on the Cell processor. OpenMP is a widely used API for parallel programming. It is attractive to support OpenMP because programmers can continue using their familiar programming model, and existing code can be re-used. We base our work on IBM's XL compiler, which already has OpenMP support for AIX multi-processor systems built with Power processors. We developed new components in the XL compiler and a new runtime library for Cell OpenMP that utilizes the Cell SDK libraries to target specific features of the new hardware platform. To describe the design of our Cell OpenMP implementation, we focus on three major issues in our system: 1) how to use the heterogeneous cores and synchronization support in the Cell to optimize OpenMP threads; 2) how to generate thread code targeting the different instruction sets of the PPE and SPE from within a compiler that takes single-source input; 3) how to implement the OpenMP memory model on the Cell memory system. We present experimental results for some SPEC OMP 2001 and NAS benchmarks to demonstrate the effectiveness of this approach. Also, we can observe detailed runtime event sequences using the visualization tool Paraver, and we use the insight into actual thread and synchronization behaviors to direct further optimizations.

1 Introduction

The Cell Broadband Engine[TM] (Cell BE) processor [8] is now commercially available in both the Sony PS3 game console and the IBM Cell Blade which represents the first product on the IBM Cell Blade roadmap. The anticipated high volumes for this non-traditional "commodity" hardware continue to make it interesting in a variety of different application spaces, ranging from the obvious multi-media and gaming domain, through the HPC space (both traditional and

B. Chapman et al. (Eds.): IWOMP 2007, LNCS 4935, pp. 65–76, 2008.

commercial), and to the potential use of Cell as a building block for very high end "supercomputing" systems [11].

This first generation Cell processor provides flexibility and performance with the inclusion of a 64-bit multi-threaded Power Processor™ Element (PPE) with two levels of globally-coherent cache and support for multiple operating systems including Linux. For additional performance, a Cell processor includes eight Synergistic Processor Elements (SPEs), each consisting of a Synergistic Processing Unit (SPU), a local memory, and a globally-coherent DMA engine. Computations are performed by 128-bit wide Single Instruction Multiple Data (SIMD) functional units. An integrated high bandwidth bus, the Element Interconnect Bus (EIB), glues together the nine processors and their ports to external memory and IO, and allows the SPUs to be used for streaming applications [9].

Data is transferred between the local memory and the DMA engine [13] in chunks of 128 bytes. The DMA engine can support up to 16 concurrent requests of up to 16K bytes originating either locally or remotely. The DMA engine is part of the globally coherent memory address space; addresses of local DMA requests are translated by a Memory Management Unit (MMU) before being sent on the bus. Bandwidth between the DMA and the EIB bus is 8 bytes per cycle in each direction. Programs interface with the DMA unit via a channel interface and may initiate blocking as well as non-blocking requests.

Programming the SPE processor is significantly enhanced by the availability of an optimizing compiler which supports SIMD intrinsic functions and automatic simdization [6]. However, programming the Cell processor, the coupled PPE and 8 SPE processors, is a much more complex task, requiring partitioning of an application to accommodate the limited local memory constraints of the SPE, parallelization across the multiple SPEs, orchestration of the data transfer through insertion of DMA commands, and compiling for two distinct ISAs. Users can directly develop the code for PPE and SPE, or introduce new language extensions [10].

In this paper we describe how our compiler manages this complexity while still enabling the significant performance potential of the machine. Our parallel implementation currently uses OpenMP APIs to guide parallelization decisions.

The remainder of the paper is laid out as follows: section two gives an overview of the compiler infrastructure upon which our work is based and presents the particular challenges of retargeting this to the novel features of the Cell platform. The next three sections of the paper look in more depth at each of these challenges and how we have addressed them, and section 6 presents some experimental results to demonstrate the benefit of our approach. We draw our conclusions in the last section.

2   System Overview

In our system, we use compiler transformations in collaboration with a runtime library to support OpenMP. The compiler translates OpenMP pragmas in the source code to intermediate code that implements the corresponding OpenMP construct. This translated code includes calls to functions in the runtimelibrary.

The runtime library functions provide basic utilities for OpenMP on the Cell processor, including thread management, work distribution, and synchronization. For each parallel construct, the compiler outlines the code segment enclosed in the parallel construct into a separate function. The compiler inserts OpenMP runtime library calls into the parent function of the outlined function. These runtime library calls will invoke the outlined functions at runtime and manage their execution.

The compiler is built upon the IBM XL compiler [1][7]. This compiler has front-ends for C/C++ and Fortran, and shares the same optimization framework across multiple source languages. The optimization framework has two components: TPO and TOBEY. Roughly, TPO is responsible for high-level and machine-independent optimizations while TOBEY is responsible for low-level and machine-specific optimizations. The XL compiler has a pre-existing OpenMP runtime library and support for OpenMP 2.0 on AIX multiprocessor systems built with Power processors. In our work targeting the Cell platform, we re-use, modify, or re-write existing code that supports OpenMP as appropriate.

We encountered several issues in our OpenMP implementation that are specific to features of the Cell processor:

- Threads and synchronization: Threads running on the PPE differ in capability and processing power from threads running on the SPEs. We design our system to use these heterogeneous threads, and to efficiently synchronize all threads using specialized hardware support provided in the Cell processor.
- Code generation: The instruction set of the PPE differs from that of the SPE. Therefore, we perform code generation and optimization for PPE code separate from SPE code. Furthermore, due to the limited size of SPE local stores, SPE code may need to be partitioned into multiple overlaid binary sections instead of generating it as a large monolithic section. Also, shared data in SPE code needs to be transferred to and from system memory using DMA, and this is done using DMA calls explicitly inserted by the compiler, or using a software caching mechanism that is part of the SPE runtime library.
- Memory model: Each SPE has a small directly accessible local store, but it needs to use DMA operations to access system memory. The Cell hardware ensures DMA transactions are coherent, but it does not provide coherence for data residing in the SPE local stores. We implement the OpenMP memory model on top of the novel Cell memory model, and ensure data in system memory is kept coherent as required by the OpenMP specification.

In the following sections, we describe how we solve these issues in our compiler and runtime library implementation.

3   Threads and Synchronization

In our system, OpenMP threads execute on both the PPE and the SPEs. The master thread is always executed on the PPE. The master thread is responsible for creating threads, distributing and scheduling work, and initializing synchronization operations. Since there is no operating system support on the SPEs, this thread also handles all OS service requests. The functions specific to the master thread align well with the Cell design of developing the PPE as a processor used to manage the activities of multiple SPE processors. Also, placing the master thread on the PPE allows smaller and simpler code for the SPE runtime library, which resides in the space-constrained SPE local store. Different from the OpenMP standard, if required by users, the PPE thread will not participate in the work for parallel loops.

Currently, we always assume a single PPE thread and specify the number of SPE threads using OMP_NUM_THREADS. Specifying the number of PPE and SPE threads separately will need an extension of the OpenMP standard. The PPE and SPE cores are heterogeneous, and there may be significant performance mismatch between a PPE thread and an SPE thread that perform the same work. Ideally, the system can be built to automatically estimate the difference in PPE and SPE performance for a given work item, and then have the runtime library appropriately adjust the amount of work assigned to different threads. We do not have such a mechanism yet, so we allow users to tune performance by specifying whether or not the PPE thread should participate in executing work items for OpenMP work-share loops or work-share sections.

We implement thread creation and synchronization using the Cell Software Development Kit (SDK) [4]. The PPE master thread creates SPE threads only when a parallel structure is first encountered at runtime. For nested parallelism, each thread in the outer parallel region sequentially executes the inner region. The PPE thread schedules tasks for all threads, using simple block scheduling for work-share loops and sections. The work sections or loop iterations are divided into as many pieces as the number of available threads, and each thread is assigned one piece. More sophisticated scheduling is left for future work.

When an SPE thread is created, it performs some initialization, and then loops waiting for task assignments from the PPE, executing those tasks, and then waiting for more tasks, until the task is to terminate. A task can be the execution of an outlined parallel region, loop or section, or performing a cache flush, or participating in barrier synchronization. There is a task queue in system memory corresponding to each thread. When the master thread assigns a task to a thread, it writes information about the task to the corresponding task queue, including details such as the task type, the lower bound and upper bound for a parallel loop, and the function pointer for an outlined code region that is to be executed. Once an SPE thread has picked up a task from the queue, it uses DMA to change the status of the task in the queue, thus informing the master thread that the queue space can be re-used.

The Cell processor provides special hardware mechanisms for efficient communication and synchronization between the multiple cores in a Cell system. The Memory Flow Controller (MFC) for each SPE has two blocking outbound mailbox queues, and one non-blocking inbound mailbox queue. These mailboxes can be used for efficient communication of 32-bit values between cores. When the master thread assigns tasks to an SPE thread, it uses the mailbox to inform the SPE of the number of tasks available for execution. Each SPE MFC also has an atomic unit that implements atomic DMA commands and provides four 128-byte cache lines that are maintained cache coherent across all processors. We use atomic DMA commands for efficient implementation of OpenMP locks, barriers[1], and cache flush operations.

4   Code Generation

Figure 1 illustrates the code generation process of the Cell OpenMP compiler. The compiler separates out each code region in the source code that corresponds to an OpenMP parallel construct (including OpenMP parallel regions, work-share loops or work-share sections, and single constructs), and outlines it into a separate function. The outlined function may take additional parameters such as the lower and the upper bounds of the loop iteration for parallel loops. In the case of parallel loops, the compiler further transforms the outlined function so that it only computes from the lower bound to the upper bound. The compiler inserts an OpenMP runtime library call into the parent function of the outlined function, and passes

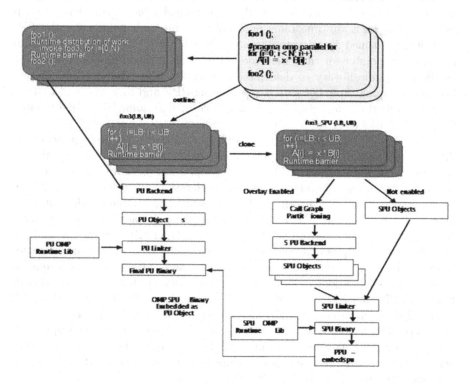

Fig. 1. Code generation process

[1] We thank Daniel Brokenshire (IBM Austin) for his implementation of barriers on Cell.

a pointer to the outlined function code into this runtime library function. During execution, the runtime function will indirectly invoke the outlined function. The compiler also inserts synchronization operations such as barriers when necessary.

Due to the heterogeneity of the Cell architecture, the outlined functions containing parallel tasks may execute on both the PPE and the SPEs. In our implementation, we clone the outlined functions so that there is one copy of the function for the PPE architecture, and one for the SPE architecture. We perform cloning during TPO link-time optimization when the global call graph is available, so we can clone the whole sub-graph for a call to an outlined function when necessary. We mark the cloned function copies as PPE and SPE procedures, respectively. In later stages of compilation, we can apply machine-dependent optimizations to these procedures based on their target architecture. Auto-simdization is one example. SPE has SIMD units that can execute operation on 128 byte data with one instruction. After cloning, the code for SPE will undergo the auto-simdization to transform scalar code into SIMD code for SPE, while the PPE has totally different SIMD instructions. In other words, we choose to clone the functions to enable more aggressive optimizations.

When a PPE runtime function in the master thread distributes parallel work to an SPE thread, it needs to tell the SPE thread what outlined function to execute. The PPE runtime function knows the function pointer for the PPE code of the outlined function to execute. However, the SPE thread needs to use the function pointer for the SPE code of the same outlined function. To enable the SPE runtime library to determine correct function pointers, the compiler builds a mapping table between corresponding PPE and SPE outlined function pointers, and the runtime looks up this table to determine SPE code pointers for parallel tasks assigned by the master thread.

At the end of TPO, procedures for different architectures are separated into different compilation units, and these compilation units are processed one at a time by the TOBEY backend. The PPE compilation units are processed as for other architectures and need no special consideration. However, if a single large SPE compilation unit is generated, it may result in SPE binary code that is too large to fit in the small SPE local store all at once. In fact, we observe this to be the case for many benchmark programs. For OpenMP, one way to mitigate this problem is to place all the code corresponding to a given parallel region in one SPE compilation unit, and generate as many SPE compilation units as there are parallel regions in the program. Using this approach, we can generate multiple SPE binaries, one for each SPE compilation unit. We can then modify the runtime library to create SPE threads using a different SPE binary on entry to each parallel region. However, there are two drawbacks to using this approach: first, an individual parallel region may still be too large to fit in SPE local store, and second, we observe through experiments that the overhead for repeatedly creating SPE threads is significantly high.

To solve the SPE code size problem, we rely on the technique of call graph partitioning and code overlay. We first partition the sub-graph of the call graph corresponding to SPE procedures into several partitions. Then we create a code overlay for each call graph partition. Code overlays share address space and do not occupy local storage at the same time. Thus, the pressure on local storage due to SPE code is greatly reduced. To partition the call graph, we weight each call graph edge by the frequency of this edge. The frequency can be obtained by either compiler static analysis or profiling. Then we apply the maximum spanning tree algorithm to the graph. Basically, we process edges in the order of their weight. If merging the two nodes of the edge does not exceed a predefined memory limitation, we merge those two nodes, update the edge weights, and continue. When the algorithm stops, each merged node represents a call graph partition comprising all the procedures whose nodes were merged into that node. Thus, the result is a set of call graph partitions. Our algorithm is a simple greedy algorithm that can be further optimized. After call graph partitions are identified, we utilize SPU code overlay support introduced in Cell SDK 2.0 and place the procedures in each call graph partition into a separate code overlay.

After the TOBEY backend generates an SPE binary (either with or without code overlays), we use a special tool called ppu-embedspu to embed the SPE binary into a PPE data object. This PPE data object is then linked into the final PPE binary together with other PPE objects and libraries. During execution, when code running on the PPE creates SPE threads, it can access the SPE binary image embedded into the PPE data object, and use this SPE image to initialize the newly created SPE threads.

5   Memory Model

OpenMP specifies a relaxed-consistency, shared-memory model. This model allows each thread to have its own temporary view of memory. A value written to or read from a variable can remain in the thread's temporary view until it is forced to share memory by an OpenMP flush operation. We find that such a memory model can be efficiently implemented on the Cell memory structure.

In the Cell processor, each SPE has just 256K directly accessible local memory for code and data. We only allocate private variables accessed in SPE code to reside in the SPE local store. Shared variables reside in system memory, and SPE code can access them through DMA operations. We use two mechanisms for DMA transfers: static buffering and compiler-controlled software cache. In both mechanisms, the global data may have a local copy in the SPE local store. The SPE thread may read and write the local copy. This approach conforms to the OpenMP relaxed memory model and takes advantage of the flexibility afforded by the model to realize memory system performance.

Some references are *regular* references from the point-of-view of our compiler optimization. These references occur within a loop, the memory addresses that they refer to can be expressed using affine expressions of loop induction variables, and the loop that contains them has no loop-carried data dependence (true, output or anti) involving these references. For such regular reference accesses to shared data, we use a temporary buffer in the SPE local store. For read references, we initialize this buffer with a DMA *get* operation before the loop executes. For write references, we copy the value from this buffer using a DMA *put* operation after the loop executes. The compiler statically generates these DMA get and put operations. The compiler also transforms the loop structure in the program to generate optimized DMA operations for references that it recognizes to be regular. Furthermore, DMA operations can be overlapped with computations by using multiple buffers. The compiler can choose the proper buffering scheme and buffer size to optimize execution time and space [9].

For irregular references to shared memory, we use a compiler-controlled software cache to read/write the data from/to system memory. The compiler replaces loads and stores using these references with instructions that explicitly look up the effective address in a directory of the software cache. If a cache line for the effective address is found in the directory (which means a cache hit), the value in the cache is used. Otherwise, it is a cache miss. For a miss, we allocate a line in the cache either by using an empty line or by replacing an existing line. Then, for a load, we issue a DMA get operation to read the data from system memory. For stores, we write the data to the cache, and maintain dirty bits to record which byte is actually modified. Later, we write the data back to system memory using a DMA put operation, either when the cache line is evicted to make space for other data, or when a cache flush is invoked in the code based on OpenMP semantics.

We configure the software cache based on the characteristics of the Cell processor. Since 4-way SIMD operations on 32-bit values are supported in the SPE and we currently use 32-bit memory addresses, we use a 4-way associative cache that performs the cache lookup in parallel. Also, we use 128-byte cache lines since DMA transfers are optimal when performed in multiples of 128 bytes and aligned on a 128- byte boundary. If only some bytes in a cache line are dirty, when the cache line is evicted or flushed, the data contained in it must be merged with system memory such that only the dirty bytes overwrite data contained in system memory. One way to achieve this is to DMA only the dirty bytes and not the entire cache line. However, this may result in small discontinuous DMA transfers, and exacerbated by the alignment requirements for DMA transfers, it can result in poor DMA performance. Instead, we use the support for atomic updates of 128-byte lines that is provided in the SPE hardware to atomically merge data in the cache line with data in the corresponding system memory, based on recorded dirty bits.

When an OpenMP flush is encountered, the compiler guarantees that all data in the static buffers in local store has been written back into memory, and that existing data in the static buffers is not used any further. The flush will also trigger the software cache to write back all data with dirty bits to system memory, and to invalidate all lines in the cache.

When an SPE thread uses DMA to get/put data from/to the system memory, it needs to know the address of the data to transfer. However, global data is linked with the PPE binary and is not directly available in SPE code. The Cell SDK provides a link-time mechanism called CESOF, that makes available to the SPE binary the addresses of all PPE global variables after these addresses have been determined. We use a facility similar to CESOF when generating SPE code.

Besides global data, an SPE thread may need to know the address of data on the PPE stack when, in source code, the procedure executing in the SPE thread is nested within a procedure executing in a PPE thread, and the SPE procedure accesses variables declared in its parent PPE procedure. Though C and Fortran do not support nested procedures (C++ and Pascal do), this case can occur when the compiler performs outlining. For example, in Figure 1, if the variable "x" were declared in the procedure that contains the parallel loop, after outlining, the declaration of "x" becomes out of the scope of the outlined function. To circumvent this problem, the compiler considers each outlined function to be nested within its parent function. The PPE runtime, assisted by compiler transformations, ensures that SPE tasks that will access PPE stack variables are provided with the system memory address of those stack variables.

6   Experimental Results

We compiled and executed some OpenMP test cases on a Cell blade that has both the PPU and the SPUs running at 3.2 GHz, and has a total of 1GB main memory. All our experiments used one Cell chip: one PPE and eight SPEs. The test cases include several simple streaming applications, as well as the standard NAS [2] and SPEC OMP 2001 [5] benchmarks. To observe detailed runtime behavior of applications, we instrumented the OpenMP runtime libraries with Paraver [3], a trace generation and visualization tool.

Figure 2 shows the PPE and SPE thread behavior for a small test case comprising of one parallel loop that is repeatedly executed one hundred times. The PPE thread is assigned no loop iterations, and is responsible only for scheduling and synchronization. The figure shows a high-level view of one complete execution of the parallel loop. The first row corresponds to the PPE thread and the remaining rows correspond to SPE threads. Blue areas represent time spent doing useful computation, yellow areas represent time spent in scheduling and communication of work items, and red areas represent time spent in synchronization and waiting for DMA operations to complete. We can clearly identify the beginning and end of one instance of a parallel loop execution from the neatly lined up set of red synchronization areas that indicate the implicit barriers being performed at parallel region boundaries, as specified by the OpenMP standard.

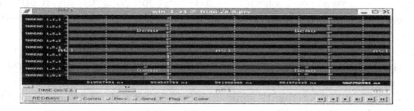

Fig. 2. PPE and SPE thread

(a) software cache only

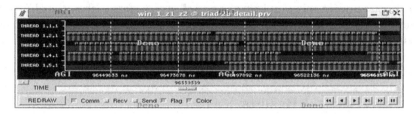

(b) optimized with static buffer

Fig. 3. Data transfer with software cache and static buffer

If we zoom into a portion of the SPE useful computation shown in Figure 2, we can see in greater detail the time actually spent in computing and the time spent in waiting for data transfers. Figure 3(a) shows this detail for a segment of execution when only the software cache is used for automatic DMA transfers. We see many areas in the SPE execution that are dominated by waiting for DMA operations to complete (red areas), and the need to optimize this application is evident. Figure 3(b) shows a similar segment of execution for the same application when it has been optimized using static buffering. We observe the improved ratio for time spent doing actual computation (blue) versus time spent waiting for DMA operations (red).

To evaluate our approach, we first tried some simple stream benchmarks, comparing the performance of code generated by our compiler with the performance of manually written code that was optimized using Cell SDK libraries and SIMD instructions. The performance comparison is shown in Figure 4. The speedups shown in this figure are the ratios of execution time when using 8 SPUs and execution time when only using the PPE. Both SIMD units and multiple cores contribute to the speedups observed. We observe that our OpenMP compiler can achieve as good a performance (except for *fft*) as the highly optimized manual code on these stream applications. *fft* performs poorly in comparison because auto-simdization cannot handle different displacements in array subscript expressions for different steps in the FFT computation. Manual code performs slightly better on *dotproduct* and *xor* because the compiler does not unroll the loop an optimal number of times.

We also experimented with some applications from the NAS and Spec OMP 2001 benchmark suites. We report speedups of the whole program normalized to one PPU and one SPU respectively in Figure 5.

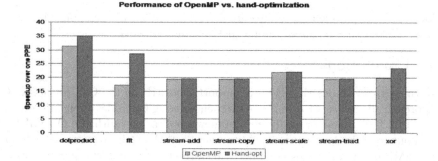

Fig. 4. Performance comparison with manually optimized code

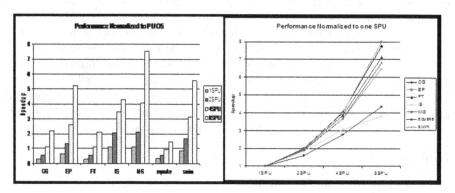

Fig. 5. Performance of benchmarks

Compared to the performance of one PPU, many applications show significant speedup with our compiler on 8 SPUs. The performance of some others is unsatisfactory, or even quite bad (such as *equake*, *CG* and *FT*). We analyzed the benchmarks and traced compiler transformations to determine the reasons for the performance shortfall. We identified two main reasons for it:

1. Limitations in our current implementation: A common reason for bad performance is the failure of static buffer optimization. Our current implementation cannot handle references complicated by loop unrolling, done either by the user or the compiler. The absence of precise alias information also prevents static buffering from being applied in many test cases. We are working on improving our compiler.
2. Applications unsuitable for the Cell architecture: *FT* and *CG* contain irregular discontinuous data references to main memory. DMA data transfers become the bottleneck in such cases.

The performance normalized to one SPU shows the scalability. Except *IS* and *equake*, all show good scalability with speedup more than 6 on 8 SPUs. *IS* does not scale up well because it performs some computation within a master pragma or critical pragma. These computations are done sequentially. For *equake*, some loops are not parallelized due to limitations of the current compiler implementation. Therefore, its speedup of 8 SPUs is only above 4.

7   Conclusions

In this paper, we describe how to support OpenMP on the Cell processor. Our approach allows users to simply reuse their existing OpenMP applications on the powerful Cell Blade, or easily develop new applications with the OpenMP API without worrying about the hardware details of the Cell processor. We support OpenMP by orchestrating compiler transformations with a runtime library that is tailored to the Cell processor. We focus on issues related to three topics: thread and synchronization, code generation, and the memory model. Our compiler is novel in that it generates a single binary that executes across multiple ISAs and multiple memory spaces.

Experiments with simple test cases demonstrate that our approach can achieve performance similar to that of manually optimized code. We also experimented with some large, complex benchmark codes. Some of these benchmarks show significant performance gains. Thus, we demonstrate that it is feasible to extract high performance on a Cell processor using the simple and easy-to-use OpenMP programming model. However, we need to further improve our compiler implementation for improved performance on a wider set of application programs.

References

1. IBM XL Compiler for the Cell BE,
   http://www.alphaworks.ibm.com/tech/cellcompiler
2. NAS parallel benchmarks,
   http://www.nas.nasa.gov/Resources/Software/npb.html
3. Paraver, http://www.cepba.upc.es/paraver
4. SDK for Cell, http://www-128.ibm.com/developerworks/power/cell
5. Spec OMP benchmarks, http://www.spec.org
6. Eichenberger, A., et al.: Vectorization for SIMD Architecture with Alignment Constraints. Programming Language Design and Implementation (PLDI) (2003)
7. Eichenberger, A., et al.: Optimizing Compiler for the Cell Processor. In: Conference on Parallel Architecture and Compiler Techniques (PACT) (2005)
8. Pham, D., et al.: The design and implementation of a first-generation cell processor. In: IEEE International Solid-State Circuits Conference (ISSCC) (February 2005)
9. Gordon, M., et al.: Exploiting Coarse-Grained Task, Data, and Pipeline Parallelism in Stream Programs. In: International Conference on Architectural Support for Programming Languages and Operating Systems (ASPLOS) (October 2006)
10. Bellens, P., et al.: CellSs: a Programming Model for the Cell BE Architecture. Supercomputing (SC) (2006)
11. Williams, S., et al.: The Potential of the Cell Processor for Scientific Computing. In: Conference on Computing Frontiers (2006)
12. Chen, T., et al.: Optimizing the use of static buffers for DMA on a CELL chip. In: Almási, G.S., Caşcaval, C., Wu, P. (eds.) KSEM 2006. LNCS, vol. 4382. Springer, Heidelberg (2007)
13. Kistler, M., Perrone, M., Petrini, F.: CELL multiprocessor communication network: Built for Speed. IEEE Micro 26(3) (May/June 2006)

CMP Cache Architecture and the OpenMP Performance

Jie Tao[1,2], Kim D. Hoàng[3], and Wolfgang Karl[3]

[1] Department of Computer Science and Technology
Jilin University, P.R. China
[2] Institut für wissenschaftliches Rechnen, Forschungszentrum Karlsruhe GmbH
Postfach 3640, 76021 Karlsruhe
[3] Institut für Technische Informatik, Universität Karlsruhe (TH)
76128 Karlsruhe, Germany
jie.tao@iwr.fzk.de

Abstract. Chip-multiprocessor (CMP) is regarded as the next generation of microprocessor architectures. For programming such machines OpenMP, a standard shared memory model, is a challenging candidate. A question arises: How to design the CMP hardware for high performance of OpenMP applications?

This work explores the answer with cache architecture as a case study. Based on a simulator, we investigate how cache organization and reconfigurability influence the parallel execution of an OpenMP program. The achieved results can direct both architecture developers to determine hardware design and the programmers to generate efficient codes.

1 Motivation

OpenMP [8] is a portable model supporting parallel programming on multiprocessor systems with a shared memory. It is gaining broader application in both research and industry. From high performance computing to consumer electronics, OpenMP has been established as an appropriate thread-level programming paradigm. In comparison to MPI [18], another dominating programming model for parallelization, OpenMP has several advantages. For example, with MPI programmers have to explicitly send messages across processes, however, OpenMP provides implicit thread communication over the shared memory. Another feature of OpenMP is that it maintains the semantic similarity between the sequential and parallel versions of a program. This enables an easy development of parallel applications without the requirement of specific knowledge in parallel programming.

Currently, with the trend of microprocessor design towards chip-multiple, OpenMP acquires increasing popularity due to its property of multithreading and the advantages mentioned above.

A chip-multiprocessor is actually a kind of SMP (Symmetric Multiprocessor) with a difference that the processor cores are integrated on a single chip. As the processors share the same memory bus, memory access penalty can be higher on this architecture than other machines like uniprocessor and SMP systems,

B. Chapman et al. (Eds.): IWOMP 2007, LNCS 4935, pp. 77–88, 2008.
© Springer-Verlag Berlin Heidelberg 2008

due to potential bus contention. In addition, more cache misses can be caused on CMPs because the per-processor cache is usually smaller than other architectures. According to an existing research work [9], up to a 4-fold of cache miss rate could be measured on a 4-processor CMP as on a uniprocessor system.

As a consequence, increasing research work addresses the design of cache architecture on CMP systems, with a current focus on the organization of the second level cache [13,16] and reconfigurable hardware [2,3]. Before the implementation of the real hardware, however, it is necessary to give a quantitative evaluation of the potential and benefit of different cache design on the running applications, here the OpenMP programs.

With this motivation, we built a cache model that simulates the whole cache hierarchy of a chip-multiprocessor system, with arbitrary level of caches, different replacement policies, a variety of prefetching schemes, and different cache line invalidation strategies. Configuration parameters, like cache size, set size, block size, and associativity, can be specified by users. Based on this simulator, we primarily explored the answer for two questions: 1) Private L2 or shared? 2) Does programs benefit from reconfiguration?

The purpose of the first study is to detect an adequate architecture of the second level cache. The performance of this cache is important because, according to the current design of CMPs, L2 is the last level of on-chip caches and an L2 miss thereby results in a reference to the main memory. We studied the impact of private caches, a standard design of SMPs, and shared caches combinedly used by all processor cores. It is found that the shared design commonly introduces better performance.

The second study targets on both static and dynamic reconfiguration of cache architectures. In the first case, the cache is specifically configured for the individual application and this configuration maintains during the whole execution of this program. A dynamic reconfigurable cache architecture, on the other hand, allows the system to adapt the cache organization to the changing access pattern of a running program. This means the cache can be reconfigured at the runtime during the execution. Our experimental results show that applications generally benefit from static reconfiguration. However, for most applications a dynamically reconfigurable cache is usually not necessary.

The remainder of the paper is organized as following. Section 2 first gives an overview of related work on simulation-based study of cache architectures. This is followed by a brief description of the developed cache model in Section 3 with a focus on its specific feature and simulation infrastructure. Section 4 presents the experimental results. The paper concludes in Section 5 with a short summary of this work.

2   Related Work

Simulation is a common used approach for evaluating hardware designs before the real hardware is implemented. A well-known example is the FLASH simulator [10] that models the complete architecture of the FLASH multiprocessor. In addition to such simulators of special purpose, tools for general research studies are also available. Representative products include SimOS [11], SIMICS [14], and SimpleScalar [1]. We choose SIMICS as an example for a deeper insight.

SIMICS is a full-system simulator with a special feature of allowing the booting of an original operating system to the simulation environment. It models shared memory multiprocessor systems and the execution of multi-threading applications. It also contains a module that models the basic functionality of caches. This module can be extended for individual purpose of different researchers. Due to this characteristics, a set of simulation-based research work in the field of cache memories are conducted on top of SIMICS. For example, Curtis-Maury et al. [4] use this simulation platform to evaluate the performance of OpenMP applications on SMP and CMP architectures in order to identify architectural bottlenecks. The L2 miss was used as a performance metric for this evaluation. Liu et al. [13] apply SIMICS to verify their proposed mechanism for implementing a kind of split organization of the second level cache.

Besides the full-system simulators, several specific cache models have also been developed. Cachegrind [19] is a cache-miss profiler that performs cache simulation and records cache performance metrics, such as number of cache misses, memory references, and instructions executed for each line of the source code. It is actually a toolkit contained in the Valgrind runtime instrumentation framework [17], a suite of simulation-based debugging and profiling tools for programs running on Linux. MemSpy [15] is a performance monitoring tool designed for helping programmers to discern memory bottlenecks. It uses cache simulation to gather detailed memory statistics and then shows frequency and reason of cache misses.

In addition, several research work use analysis models to predict the cache access behavior and achieve performance metrics to evaluate either the hardware or the application. StatShare [7] is such an example, which predicts the cache miss rate based on the reuse distance of memory references.

Overall, cache architecture has been studied using simulation in the last years and chip-multiprocessor has also been addressed. We follow this traditional approach, but conduct a novel investigation, i.e. studying the adaptivity between OpenMP performance and the cache organization.

3   The Cache Simulator

As available simulators do not deliver all required properties, e.g. performance metrics, for this specific study, we developed a cache simulator called CASTOR (CAche SimulaTOR).

CASTOR Features. First, CASTOR simulates the basic functionality of the cache hierarchy on a CMP machine. For flexibility, cache-associated parameters such as size, associativity, cache level, and write policies can be arbitrarily specified by users. Corresponding to most of the realistic cases, we assume that one processor runs only a single thread and the master thread, which is responsible for the sequential part of an application, is mapped to the first processor.

CASTOR simulates several cache coherence protocols for filling the different requirement of users. In addition to the conventional MESI scheme, other protocols like MSI and MEOSI are also supported. These protocols differ in the number of status of a cache line, where MESI scheme uses four status, Modified, Exclusive (unmodified), Shared (unmodified), and Invalid, to identify a cache line, while MSI applies only three of them and MEOSI deploys an additional one "Owner" to show the home node of the cached data. False sharing scenarios are identified in order to make this simulator more general that it can also be applied for other research work, for example, code optimization. For the same reason, mechanisms of classifying cache misses and various prefetching schemes are implemented within CASTOR.

Simulation Infrastructure. CASTOR requires a frontend to deliver the runtime memory references of an OpenMP program. We could integrate CASTOR into the SIMICS environment, however, for this research work, a complicated full-system simulator is not necessary. In this case, we deployed Valgrind to provide the needed information and built an interface between CASTOR and this frontend for an on-line simulation. Alternatively, memory references can be stored in a trace file enabling simulations with different cache parameters, without having to run the application again.

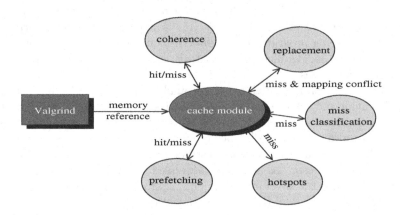

Fig. 1. CASTOR overall infrastructure

CASTOR, on the top level, is composed of several modules. As shown in Figure 1, the *cache module*, the main component of this infrastructure, is responsible for the basic functionality of a cache hierarchy. For each memory access from Valgrind it performs the search process in the caches available for the thread issuing this reference and thereby determines whether the access is a hit or a miss at each cache level. During this process, other modules could be initiated for additional functions. First, the *coherence module* is called to handle the cache coherence issues. In case of a mapping conflict, the *replacement module* is deployed to achieve a free cache line for incoming data. For a cache miss, the *miss classification module* is adopted to determine the miss type, which lies in four categories: compulsory, capacity, conflict, and invalidation miss. Compulsory miss occur when the data is first accessed, hence this kind of miss is also called first reference or cold miss. In this case, the data is still in the main memory and has to be loaded to the cache. Conflict misses occur when a data block has to be removed from the cache due to mapping overlaps. Capacity misses occur when the cache size is smaller than the working set size. The information about the miss type can give the programmers valuable help in code optimization. The *hotspot module* is responsible for mapping the cache miss to the functions in the source program. This feature allows CASTOR to deliver simulation results at high-level and also the users to find access bottlenecks. Finally, the *prefetching module* is called when the user has specified a kind of prefetching. In this case, appropriate data blocks are loaded to the cache and the cache state is correspondingly updated.

Performance Data. CASTOR provides at the end of simulation a set of performance metrics, such as statistics on cache hits, cache misses and each miss category, cache line invalidations, false sharing, and prefetching. Performance information can be given at the basis of complete program or for each individual function contained in a program.

4   Experimental Results

Based on CASTOR, we could study the influence of different CMP cache organization on the performance of OpenMP applications. Six applications from both the SPEC [6] (Equake, Mgrid, and Gafort) and NAS [5,12] (CG, FT, and EP) OpenMP benchmark suite are tested. For smaller memory access traces the SPEC applications are simulated with the test data size and applications from the NAS Parallel Benchmark are configured with CLASS S.

Private or Shared L2? The first study addresses the organization of the cache level closest to the main memory, i.e. the L2 cache. Applications are executed on systems with 2, 4, and 8 cores separately and the number of cache misses are measured. To allow a fair comparison, the size of the shared L2 cache is the sum of all local L2s in the private case. Cache parameters are chosen as: 32 bytes of cache line, 4-way write-through L1 of 16K, 8-way write-back L2 of 512K (private), MESI coherence protocol, and the LRU replacement policy that chooses the least recently used cache block as candidate to store the requested data . We select several interesting results for demonstration.

Figure 2 is the experimental result with the CG code. For a better understanding of the cache behavior, we show the miss in different classifications rather than only depicting the total misses which correspond to the height of the columns in the figure. For this figure and all others in the following the number of misses is shown in a unit of one million.

As can be seen in the left diagram, a shared cache performs generally better than the private caches. In the case of 4 and 8 processors, almost all misses that occur with private cache are eliminated. This outstanding performance could

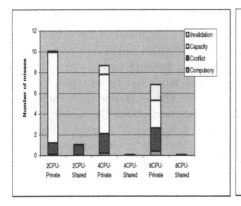
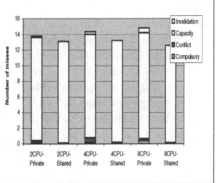

Fig. 2. L2 misses of the CG application

be resulted by the relative large cache size, in contrast to the working set, and perhaps does not suffice for a general conclusion. Therefore, we simulated the code again with a smaller L2 of 64K. As shown in the right diagram of Figure 2, the same result can be observed: shared L2 outperforms private caches.

Observing both diagrams, it can be seen that for the CG program capacity miss is the main reason of poor performance of private caches. It can also be seen, more clearly in the left diagram, that the performance gain achieved with shared L2 is also primarily due to the reduction of the capacity miss. This can be explained by the fact that in the shared case a larger L2 (in size of combined private L2s) is available to each processor and as a consequence capacity problem can be relaxed. However, the CG application also shows a reduction of conflict miss in the shared L2 cache. This feature depends on the access pattern of this program. As the data requested by a single processor has different mapping behavior in shared and private caches, conflicts that occur in a private cache could be removed with a larger shared cache.

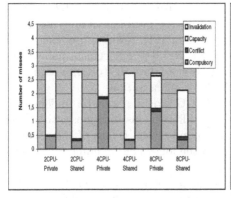
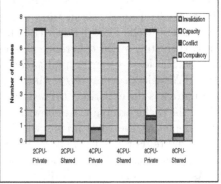

Fig. 3. L2 misses of the FT (left) and Equake (right) applications

Similar to CG, other applications also prefer to a shared L2. Figure 3 demonstrates the result with the FT and Equake program. For both codes, it can be seen that the better performance with a shared L2 is contributed by the reduction of compulsory misses. The capacity miss, another dominant miss category with these programs, however, can not be removed with a shared cache. This means the combination of the L2 caches on all processors can not achieve a cache memory that is large enough to hold the needed data for running these programs.

Overall, the experimental results depict that OpenMP applications benefit from the shared cache. Although different codes have their own contribution to this, several common reason exist. For example, parallel threads executing a program share partially the working set. This shared data can be reused by other processors after being loaded to the cache, reducing hence the compulsory miss. Shared data also saves space because it is only cached once. This helps to reduce the capacity miss. In addition, there are no invalidation misses with a shared cache.

Static Reconfigurable Cache? The second experiment aims at finding how the performance of an application varies across different cache configurations. For this we simulated the cache behavior of each application with various cache size, line size, and associativity. Number of total misses in the L1 cache were measured.

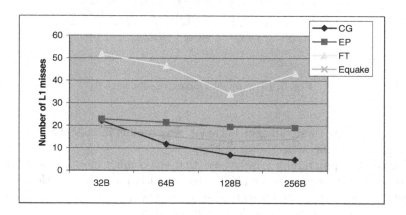

Fig. 4. L1 miss with changing block size

Figure 4 is the result with changing cache block size. It can be seen that applications differ in cache access behavior with this parameter. CG presents the best scalability, where cache miss keeps reducing as the block size increases. EP shows the same behavior, but with only slight reduction in cache miss. The other programs, FT and Equake achieve the best performance with a cache block size of 128B and further enlarging the blocks worsens cache performance.

To better understand these different behavior, we have examined the miss classification of two representative programs: CG and FT. We found that the good performance and scalability of CG is mainly contributed by the reduction of capacity miss, but also the conflict miss which maintains reducing even though only slightly. For the FT code, however, the conflict miss grows with 256B and the reduction in capacity miss can not compensate for this. As a consequence, more misses are caused with a 256B cache than a 128B one.

For changing associativity, nevertheless, the result is not so identical. As can be seen from Figure 5, FT and EP have increasing cache misses with larger associativity; CG behaves similarly with various set size; Equake benefits significantly from the increasing associativity; and the rest two applications, Gafort and Mgrid, have a better performance with a 4-way cache but a further increment in associativity does not introduce more cache hits.

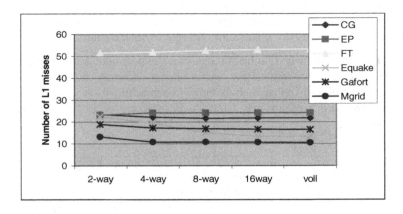

Fig. 5. L1 miss with changing associativity

For explaining these different behavior, we examine further the miss classifications of all applications shown in Figure 6.

For the CG program, it can be seen that the conflict miss, as expected, decreases with the increasing number of blocks within a cache set. However, the capacity miss, another dominating miss of this code, grows up with the associativity. Combined, the cache performance gets worse.

For the FT program, nevertheless, conflict misses increase with the associativity. This behavior is not expected. Hence, we have examined the three primary individual functions in this program and observed that one of them has less miss with larger cache sets but the others perform best with a 2-way cache. This means an individual tuning with respect to the functions, e.g. using a reconfigurable cache architecture capable of changing the configuration at the runtime, could be needed by this application.

EP shows less miss with a 4-way cache than a 2-way cache, but a further increase in associativity introduces more cache misses. It can be observed that EP has capacity problem. However, the capacity miss does not change across different set size. This specific behavior lies singly on the conflict miss and can only be explained by the individual access pattern of this application.

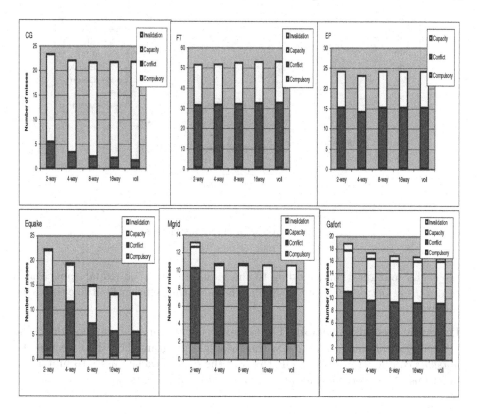

Fig. 6. Miss classification with changing associativity

Equake achieves the best performance with the changing associativity: number of cache misses keeps going down up to full-associative caches. As shown in Figure 6, this is completely contributed by the reduction of conflict miss. Hence, increasing the blocks of a cache set helps this application to tackle the mapping conflict.

Mgrid and Gafort show a significant miss reduction when the associativity switched from 2-way to 4-way. However, further enlarging the cache set does not introduce much change in cache performance. This means for both applications a 4-way cache is preferred.

For changing cache size, as shown in Figure 7, no worse behavior can be seen with the increase of the cache capacity. Applications differ in this behavior in that some benefit more and others less. For example, CG needs a large cache for a high cache ratio, but for FT a larger cache can not remove the misses significantly.

In summary, OpenMP applications have different requirement in cache organization. For example, a specific line size can significantly improve the cache performance of an individual application, but introduces no gain with another program. A large cache can potentially considerately reduce the cache miss of one application, but is not necessary for another. This means a static tuning of

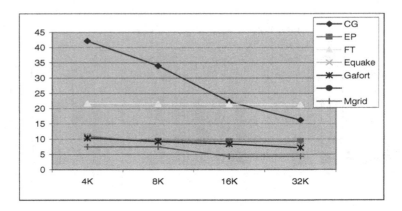

Fig. 7. L1 miss with changing cache size

the cache configuration according to the requirement of individual applications can result in better performance and at the same time efficiently utilize the cache resource.

Dynamic Reconfiguration? The last experiment aims at examining the cache behavior of different functions within a single program. For this, we use the *hotspots* component of CASTOR to order the overall cache misses to individual functions. Again we measured the number of total L1 misses.

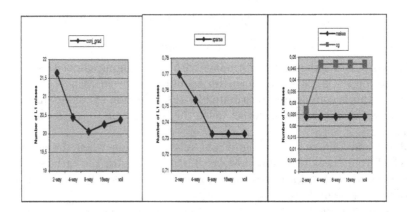

Fig. 8. L1 miss of functions with changing associativity

Figure 8 is a sample result achieved with the FT code. Three diagrams in the figure depict the changing behavior, with respect to the cache associativity, of the four functions with the most cache misses. It can be seen that the best set size for both *conj_grad()* and *sparse()* is 8, while for *makea()* a 2-way cache is better and *cg()* has no specific need. This indicates that a dynamic tuning of the cache line size potentially improves the performance of CG. Similar result has also been acquired with FT. However, for other applications and cache parameters, no considerate distinction has been observed. Therefore, we conclude that a runtime cache reconfiguration is usually not necessary for the OpenMP execution.

5   Conclusions

Chip Multiprocessor is the trend of microprocessor design and OpenMP is an appropriate programming model for it. As the memory wall keeps widening, cache performance becomes increasingly important, especially for chip multiprocessors with potentially more cache misses.

This paper investigates the impact of various cache designs on the performance of OpenMP programs on chip multiprocessor systems. The study is based on a self-written cache simulator that comprehensively models the cache functionality and provides thereby performance metrics for evaluating the cache access behavior. We have focused on the structure of the second level cache and issues with reconfigurable caches. The achieved results can direct both hardware developers and programmers to design efficient architecture and applications.

References

1. Austin, T., Larson, E., Ernst, D.: SimpleScalar: An Infrastructure for Computer System Modeling. Computer 35(2), 59–67 (2002)
2. Benitez, D., Moure, J.C., Rexachs, D.I., Luque, E.: Evaluation of the Field-programmable Cache: Performance and Energy Consumption. In: Proceedings of the 3rd conference on Computing frontiers (CF 2006), Ischia, Italy, May 2006, pp. 361–372 (2006)
3. Carvalho, M.B., Goes, L., Martins, C.: Dynamically Reconfigurable Cache Architecture Using Adaptive Block Allocation Policy. In: Proceedings of the 20th International Parallel and Distributed Processing Symposium (IPDPS) (April 2006)
4. Curtis-Maury, M., Ding, X., Antonopoulos, C., Nikolopoulos, D.: An Evaluation of OpenMP on Current and Emerging Multithreaded/Multicore Processors. In: Proceedings of the First International Workshop on OpenMP (IWOMP), Eugene, Oregon USA (June 2005)
5. Bailey, D., et al.: The NAS Parallel Benchmarks. Technical Report RNR-94-007, Department of Mathematics and Computer Science, Emory University (March 1994)
6. Saito, H., et al.: Large System Performance of SPEC OMP 2001 Benchmarks. In: Zima, H.P., Joe, K., Sato, M., Seo, Y., Shimasaki, M. (eds.) ISHPC 2002. LNCS, vol. 2327, pp. 370–379. Springer, Heidelberg (2002)
7. Petoumenos, P., et al.: Modeling Cache Sharing on Chip Multiprocessor Architectures. In: Proceedings of the 2006 IEEE International Symposium of Workload Characterization (2006)
8. Chandra, R., et al.: Parallel Programming in OpenMP. Morgan Kaufmann, San Francisco (2000)
9. Fung, S.: Improving Cache Locality for Thread-Level Speculation. Master's thesis, University of Toronto (2005)

10. Gibson, J., Kunz, R., Ofelt, D., Horowitz, M., Hennessy, J., Heinrich, M.: FLASH vs (Simulated) FLASH: Closing the Simulation Loop. In: Proceedings of the 9th International Conference on Architectural Support for Programming Languages and Operating Systems (ASPLOS), November 2000, pp. 49–55 (2000)
11. Herrod, S.A.: Using Complete Machine Simulation to Understand Computer System Behavior. PhD thesis, Stanford University (February 1998)
12. Jin, H., Frumkin, M., Yan, J.: The OpenMP Implementation of NAS Parallel Benchmarks and Its Performance. Technical Report NAS-99-011, NASA Ames Research Center (October 1999)
13. Liu, C., Sivasubramaniam, A., Kandemir, M.: Organizing the Last Line of Defense Before Hitting the Memory Wall for CMPs. In: Proceedings of the International Symposium on High-Performance Computer Architecture (HPCA 2004), Madrid, Spain, February 2004, pp. 176–185 (2004)
14. Magnusson, P.S., Werner, B.: Efficient Memory Simulation in SimICS. In: Proceedings of the 8th Annual Simulation Symposium, Phoenix, Arizona, USA (April 1995)
15. Martonosi, M., Gupta, A., Anderson, T.: Tuning Memory Performance of Sequential and Parallel Programs. Computer 28(4), 32–40 (1995)
16. Molnos, A.M., Cotofana, S.D., Heijligers, M.J.M., van Eijndhoven, J.T.J.: Static Cache Partitioning Robustness Analysis for Embedded On-chip Multi-processors. In: Proceedings of the 3rd conference on Computing frontiers (CF 2006), Ischia, Italy, May 2006, pp. 353–360 (2006)
17. Nethercote, N., Seward, J.: Valgrind: A Program Supervision Framework. In: Proceedings of the Third Workshop on Runtime Verification (RV 2003), Boulder, Colorado, USA (July 2003), http://developer.kde.org/~sewardj
18. Pacheco, P.: Parallel Programming with MPI. Morgan Kaufmann, San Francisco (1996)
19. WWW. Cachegrind: a Cache-miss Profiler, http://developer.kde.org/~sewardj/docs-2.2.0/cg_main.html#cg-top

Exploiting Loop-Level Parallelism for SIMD Arrays Using *OpenMP*

Con Bradley and Benedict R. Gaster

ClearSpeed Technology Plc
3110 Great Western Court
Hunts Ground Road
Bristol BS34 8HP UK
{ceb,brg}@clearspeed.com

Abstract. Programming SIMD arrays in languages such as C or FORTRAN is difficult and although work on automatic parallelizing programs has achieved much, it is far from satisfactory. In particular, almost all 'fully' automatic parallelizing compilers place fundamental restrictions on the input language.

Alternatively *OpenMP* provides an approach to parallel programming that supports incremental improvements to applications that places restrictions in the context of preserving semantics of a parallelized section. *OpenMP* is limited to a thread based model that does not define a mapping for other parallel programming models.

In this paper, we describe an alternative approach to programming SIMD machines using an extended subset of *OpenMP(OpenMP SIMD)*, allowing us to model naturally the programming of arbitrary sized SIMD arrays while retaining the semantic completeness of both *OpenMP* and the parent languages.

Using the CSX architecture we show how *OpenMP SIMD* can be implemented and discuss future extensions to both the language and SIMD architectures to better support explicit parallel programming.

1 Introduction

With the rise of both SIMD processors, such as ClearSpeed's SIMD Array Processor [1], and the current unstoppable trend to put more and more cores on to a single die, parallel programming is back with a bang! For anyone who has tried to write a reasonably large parallel program then this will be seen with at least a slight frown. Why? In a nutshell parallel programming is hard and a choice must be made as to which of the many parallel programming languages to use.

While there is no magic wand to wave that can easily resolve the problem of parallel programming, there has been a number of recent proposals for programming languages, including Brook [2], Cilk [3], and C^n [4], and a number of simple language extensions for parallel programming, including *OpenMP* [5] and X10 [6]. Other than Brook, which introduces a programming model based on Streams, C^n, *OpenMP*, and X10 are all based upon existing programming

B. Chapman et al. (Eds.): IWOMP 2007, LNCS 4935, pp. 89–100, 2008.

languages C^1 [7] and Java [8], respectively, and as such provide parallel programming in the context of well understood and widely used programming paradigms. It is only *OpenMP* that has seen wide adoption in the marketplace and is the only major parallel programming extension to C that is supported by the major compiler vendors, including Intel (Intel C Compiler) and Microsoft (Visual Studio).

OpenMP as defined is intended as a light-weight approach to programming threads. In this paper, we describe an alternative interpretation of *OpenMP*, called *OpenMP SIMD*, that rather than mapping to the standard runtime parallelism of threads looks instead at how it can be deployed for programming SIMD machines. In particular, we show that the loop level parallelism described by *OpenMP* maps well to data-parallelism as an alternative to the runtime approach described traditionally by *OpenMP*.

While we believe this alternative semantics for *OpenMP* is valid for any SIMD architecture, including Intel's Streaming SIMD Extensions [9], we focus here on programming the CSX architecture [1]. The CSX architecture is a SIMD array with a control unit and a number of processing elements (PE), with each PE having a relatively small private storage and supports an asynchronous I/O mechanism to access larger shared memory.

To date the CSX architecture would have been programmed using the explicit parallel programming language C^n but such an approach has some draw backs:

- Any code must be completely ported to use explicit parallelism before it can be run; and
- Users have large amounts of C and *OpenMP* code that they do not want to re-write.

A mapping from *OpenMP* to either C^n or directly to the CSX architecture provides an alternative approach as *OpenMP* provides a piecemeal approach to parallelizing a given program and by definition supports any legacy C and *OpenMP* code.

The remaining sections of this paper are as follows; Section 2 introduces the CSX architecture, including how it is intended to be programmed; Section 3 is broken into a number of sub-sections describing the translation of *OpenMP SIMD* to the CSX architecture; Section 4 gives two examples of *OpenMP SIMD* in practice, highlighting both the translation and the small but necessary extensions to *OpenMP*; and finally, Section 5 concludes.

2   The CSX Architecture

The CSX family of co-processors [1] are based around a multi-threaded array processor (MTAP) core optimized for high performance, floating point intensive applications. The most recent product includes an MTAP processor, DDR2 DRAM interface, on-chip SRAM and high speed inter-processor I/O ports.

---

[1] *OpenMP* provides models for C++ and FORTRAN but we choose to focus on the C++ variant.

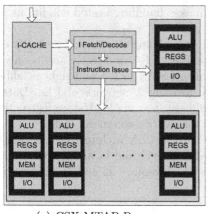

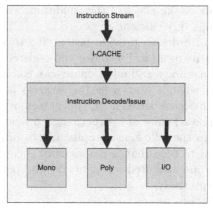

(a) CSX MTAP Processor   (b) Instruction Execution

Fig. 1. CSX Architecture

Figure 1(a), visualizes the MTAP architecture showing a standard, RISC-like, control unit with instruction fetch, caches, and I/O mechanisms. A single instruction stream comprising of fairly standard, three operand instructions, is made up of mono (non-parallel) instructions executed by a single RISC-style processor and SIMD instructions (poly) executed, in lock-step, by a number of poly execution (PE) cores[2].

As noted above the processor executes a single instruction steam; each instruction is sent to one of the functional units, which may be in the mono, or poly execution units or one of the I/O controllers. Figure 1(b) shows this visually.

The main change from programming a standard processor is the concept of operating on parallel data. Data to be processed is assigned to 'variables' which have an instance on every PE core and are operated on in parallel: we call this poly data. Intuitively, this can be thought of as a data vector which is distributed across the PE core array.

Variables which only need to have a single instance (e.g. loop control variables) are known as mono variables; they behave exactly like normal variables on a sequential processor and can be 'broadcast' to all PEs if required in the poly domain.

Code for the mono and poly execution units is handled similarly, except the handling of conditional code execution. The mono unit uses conditional jumps to branch around code like a standard RISC architecture. This affects both mono and poly operations. The poly unit uses an enable register to control execution of each PE. If one or more of the bits of that PE enable register is zero, then the PE core is disabled and most instructions it receives will perform as though a *NOP* has been issued. The enable register is a stack, and a new bit, specifying the result of a test, can be pushed onto the top of the stack allowing nested predicated execution. The bit can later be popped from the top of the stack to

---

[2] The current CSX600 processor has 96 PEs.

remove the effect of the condition. This makes handling nested conditions and loops very efficient.

In order to provide fast access to the data being processed, each PE core has its own local memory and register file. Each PE can directly access only its own storage and additionally can communicate with its left and right neighbors via what is known as the swazzle path, forming a network of PEs. Data is transferred between PE (poly) memory and registers via load/store instructions while a family of instructions is provided for swazzle communication.

In the following mapping of *OpenMP* to the CSX architecture it is important to note that mono memory (i.e. card memory) and local PE (poly) memory are allocated separate address spaces which require 32 and 16 bit pointers, respectively.

3 *OpenMP SIMD*

As discussed in the introduction we are looking at using *OpenMP* to exploit loop level parallelism and as such the translation described below supports only one parallel construct, namely, the work-sharing loop. For this construct, we retain the fork-join 'threading' model. The master thread runs on the mono processor and is responsible for creating the 'thread' gang to execute the iterations of the work-sharing loop. Where the translation differs from the multi-core threading model traditionally associated with *OpenMP* is that each 'thread' is modeled as a PE element. At the end of the loop there is an implicit barrier. *OpenMP* expects the master thread to participate in the loop calculations but in our model this is not the case due to the asymmetry between the PE and mono processor architecture. We assume that all housekeeping operations are carried out on the mono processor.

3.1 Work-Sharing Constructs

The *OpenMP SIMD* translation actually provides two alternative mappings descrined in the following to sections.

Poly Loop Translation. *OpenMP* states that the index variable of the loop to be parallelized becomes private for the duration and thus a straightforward translation of the *OpenMP* loop pragma to *OpenMP SIMD* is to make the control variable poly and to increment by the number of PEs each time around the loop. Such a translation has the effect of unrolling the loop by the number of PEs. Mapping the test expression of the loop to use the poly control variable and increment with the defined increment plus the number of PEs in the final clause of the loop definition, provides all that is needed for the loop control definition. All that is remaining is to translate the body of the loop. The mapping is defined as follows:

```
#pragma omp parallel for clause clause ...
  for (i = e₁;  i op e₂;  i = i opₛ incr) { stmts; }
```

is translated to[3]:

```
{ poly int iₚ = e₁;
    for (iₚ = iₚ opₛ get_penum() * incr; iₚ op e₂; s₁ₚ )
      { stmtsₚ; }
}
```

Here *OpenMP* requires that *op* is one of $<$, $>$, $\leq$, or $\geq$ and op_s is $+$ or $-$. With this s_{1_p} becomes[4] $i_p = i_p \; ops \; (incr \; * \; \_NUM\_PES\_)$.

Finally, we must provide a translation for $stmts_p$. Firstly, all automatics defined locally to the scope of the loop become poly, i.e. privates in *OpenMP*, and any reference to the index variable i is replaced with i_p. Secondly, the translation must replace all accesses to arrays indexed by newly introduced poly expressions with poly temporaries, and insert calls to runtime routines that copy data between memory spaces, accounting for the strided access implied by parallelizing by the number of PEs. For example, consider the following loop statement which uses i to index into an array of mono doubles:

```
double tmp = V[i] ;
```

In the first part of the translation this becomes:

```
poly double tmpₚ = V[iₚ] ;
```

Assuming that V is defined globally and thus in *OpenMP* terminology shared, then this expression becomes the list of statements[5]:

```
poly double tmpₚ;
double * poly Vₚ = ((double * poly) V) + iₚ;
memcpym2p(&tmpₚ, Vₚ, sizeof(double));
```

where $memcpym2p$ is one of a family of memory transfer routines forming part of the *OpenMP SIMD* runtime, with this routine in particular performing data transfers from mono memory into local PE memory.

Mono Loop Translation. The poly loop translation given above is both obvious and straightforward but in practice may not be the most effective approach to implementing the loop on a SIMD style architecture.

To illustrate this consider the simple poly loop[6]:

```
for (iₚ = 0; iₚ < bound; iₚ++) { ... }
```

As all control flow is handled in the mono processor a poly loop must be implemented by evaluating the conditional $i_p < bound$ such that all, some, or none of the PEs are disabled. Then a mono instruction is issued to determine if all of the PEs are disabled and if so then exit the loop, otherwise continue with

---

[3] The *OpenMP SIMD* runtime function poly int $get\_penum(void)$ returns the unique number assigned to each PE, given in the sequence 0..95.

[4] $\_NUM\_PES\_$ is defined statically as the number of PEs for a given implementation.

[5] The type expression *double *poly* should be read as a poly pointer to a mono double.

[6] Assuming *bound* to be mono.

the next iteration. The overhead of checking wheather the PEs are disabled or not is expensive and is often not necessary, as a mono loop can be used instead.

To explain the mono loop translation we consider again the loop from the previous section:

```
#pragma omp parallel for clause clause ...
for (i = e₁; i op e₂; i = i op incr) { stmts; }
```

which can be translated into the mono loop:

$$\text{for } (i = e_1; \ i \ op \ (e_2 - (e_2 \% \_\_NUM\_PES\_\_)); \\ i = i \ op \ (incr \ * \ \_\_NUM\_PES\_\_)) \\ \{ \ stmts_p \ \}$$

Finally a translation for $stmts_p$ is required, which is the same as in the previous section except in the case when the control variable i is referenced. In this case the translation needs to account for the fact that $OpenMP$, and thus the translation, requires that for the duration of the loop i is marshaled as a private and to account for the unrolling factor the PE number must be added into any expression referencing i. Again considering the loop defined statement:

```
double tmp = V[i];
```

the first part of the translation gives:

```
poly double tmpₚ = V[i];
```

As i must be treated as a private, accounting for the stride due to loop unrolling, the final translated code becomes:

```
poly double tmpₚ;
double * poly Vₚ = ((double * poly) V) +
                                    (i + get_penum() * incr);
memcpym2p(&tmpₚ, Vₚ, sizeof(double));
```

If one is not careful it is easy to think at this point that the translation for mono loops provided above is enough, but we must consider the case when the number of PEs is not a multiple of the number of PEs. To see that there may in fact be a problem, consider again the translated test condition for the loop $i \ op \ (e_4 - (e_4 \% \_\_NUM\_PES\_\_))$.

The expression $(e4 - (e_4 \% \_\_NUM\_PES\_\_))$ always results in an expression that is a multiple of the number of PEs. It is straightforward to see that any remaining number of iterations will not be performed by the loop, and as is the case with optimizations like software-pipelining [10] the mono loop translation requires some additional post loop code to account for these missing iterations. This is easily achieved by augmenting the translation with the post loop code:

$$\text{if } (get\_penum() < (e_4 \% \_\_NUM\_PES\_\_)) \ \{ \ stmts_p; \ \}$$

Where $stmts_p$ is the same set of statements from the translated loop body and the conditional simply disables any PEs not required for the remaining calculation. An important consideration is to note that as the conditional is poly then $stmts_p$ will always be executed. If the condition $get\_penum() <$ $e_4\%\_NUM\_PES\_$ evaluates to false this could be expensive depending on the size of the sequence $stmts_p$. This overhead can be easily avoided by adding in an additional mono conditional of the form:

```
if (e4%_NUMpES_)
```

3.2 Data Environment

$OpenMP$ defines a set of constructs to describe how data is shared in parallel regions and we must capture how these are mapped into $OpenMP\ SIMD$.

$OpenMP$ defines storage as being either private to a thread or shared, i.e. globally shared for all threads. The obvious translation into $OpenMP\ SIMD$ is to map private variables to PE memory (poly) and shared to mono memory.

The remainder of this section considers the consequences of this simple translation and provides some extensions to $OpenMP$ when the mapping does not quite capture the required semantics.

Sharing Attribute Rules. $OpenMP$ defines a set of rules for describing sharing attributes of variables within parallel constructs. These are variables without explicit sharing attributes, i.e. the default rules, and we need to confirm their semantics for $OpenMP\ SIMD$. These rules apply in two cases:

1. Variables explicitly declared within the work sharing loop. For our translation the default rules work well and do not need modification.
2. Variables within called functions. $OpenMP$ refers to these as being within a region but not a construct, meaning that syntactically a variable does not appear in a parallel construct but at runtime the function is called from a parallel region. The rules for data sharing are as in the explicit case but with one major addition: Formal arguments passed by reference inherit the data sharing attributes of the actual parameters.

This implies that at compile time, it is impossible for a compiler to know if a function will be called from a parallel region and thus what the data sharing attributes of its formal parameters will be. In order for the translation to $OpenMP\ SIMD$ to work efficiently the user is required to provide this information. This leads to the introduction of one of three extensions to $OpenMP$ as it stands:

```
#pragma omp private_function <name>
```

This $private\_function$ clause marks the function $< name >$ as being called from a parallel region and hence the default data sharing rules are to be applied. The default is that all functions are serial, i.e. mono, but this can be overridden via a default clause. In a typical program, parallel functions can be grouped thus reducing the amount of explicit clauses needed.

Unlike standard $OpenMP$ the formal arguments to a function for the CSX architecture can be either mono or poly but not both. This leads to the observation that data-sharing attributes of formal parameters need to be specified at compile time as they affect code generation. This is captured in *OpenMP SIMD* by introducing a second extension to *OpenMP*:

```
#pragma omp param <name> (list)
```

Where list is defined as a comma separated list, one entry per formal parameter to $< name >$; each entry can be of the form:

- private - formal is private (i.e. poly)
- shared - formal is shared (i.e mono)

Variable argument parameters and function pointers are not supported.

The final extension needed for the translation is analogous to that for a function's formal arguments but applies to the return value as it can be either mono or poly but not both:

```
#pragma omp private_return <name>
```

This *private_return* clause marks the function $< name >$ as having a private return value, i.e. poly. The default is that functions return a shared value, i.e. mono.

Any function can have any number of sharing directives applied to describe all possible calling conditions. This defines a set of 'signatures' describing the set of possible calling conditions. Given these extra annotations a compiler might choose to use a C++ [11] style name mangling scheme at function definition time and at function call sites to generate the required function signature. For example, a simple void function, void foo(void), with no parameters which can be called from both a parallel and non-parallel region could have two instances, *_POLY_foo* and *_MONO_foo*. In a serial context, the mono version is called, and in a poly context the poly version is called. The linker will then select the appropriate versions. This scheme is also applied to a function's formal arguments to generate C++ style mangled names.

It is important to note that for this system to be sound, it must at all times be known if a piece of code is in a parallel region or not. This is where *OpenMP SIMD* differs slightly from *OpenMP* as it requires this to be enforced by the user through application of the above extensions.

Data-Sharing Attribute Clauses. *OpenMP* assumes that multiple threads execute within the same shared address space and thus inter-thread communication is straightforward. Data is sent to other threads by assigning to shared variables. *OpenMP* also provides support for a variable to be designated as private to each thread rather than shared, causing each thread to get a private copy of a particular variable for the duration of the parallel construct.

This notion of shared and private fits well within our SIMD model of OpenMP, producing the mapping of *private* to *poly* and *shared* to *mono*. The *firstprivate* and *lastprivate* clauses are also supported with extra runtime functions provided to perform the initialization and finalization.

The reduction clause describes the application of a reduction operator to be applied to a list of private variables. The reduction operator comes from the base language and is applied to each (private) copy of the variable at the end of the parallel construct. This construct is supported by introducing a private copy of the reduction variable, initialized to the particular reduction operator's identity, at the start of the parallel construct. The reduction is then preformed across all PEs. The swazzle path is used to reduce into either the top or bottom PE, where the data can be quickly transferred into mono space and assigned to the original reduction variable. For example, consider the following reduction clause and loop:

```
#pragma omp parallel for reduction(+:acc)
for (...) { ... ;   acc += ...;   }
```

which becomes[7]:

```
poly double acc_p = 0.0;
for (...) { ... ;   acc_p += ... ; }
acc += reduce_sum(acc_p);
```

4   An Example

As an example of *OpenMP SIMD* we consider performing a Monte-Carlo simulation to calculate European option prices [12]. The intention here is to show the extensions in practice. Due to space restrictions we choose not to show the translated code but rather concentrate our discussion of the extensions themselves.

Monte-Carlo simulation problems include derivative pricing, portfolio risk management and Markov chain simulation. In financial modeling, the problem of finding the arbitrage-free value of a particular derivative can be expressed as the calculation of a given integral. As the number of dimensions grows, today's financial mathematicians use Monte-Carlo simulation methods to solve problems in a tractable manner.

Monte-Carlo simulations can be easily parallelized, since tasks are largely independent and can be partitioned among different processors, and the need for communication is limited. Due to the simple but highly computational nature of Monte-Carlo it is clear that any performance increase in their implementation can lead to substantial speed improvements for the entire system and significant advances in time to result. The following function implements Monte-Carlo option pricing in *OpenMP SIMD*:

```
void MCOption(double S, double E, double r,
              double  sigma, double  T,
          int nSimulations,
          double *Pmean, double *width) {
```

---

[7] The function *double reduce_sum(poly double)* is one of a family of reduction operators provided as part of the *OpenMP SIMD* runtime, with this function performing a double precision sum across the PEs placing the scalar result into mono.

```
double sumx, sumx2, tmp1, tmp2, stdev;
int i;

sumx=0; sumx2=0;
tmp1=(r-0.5*sigma*sigma)*T;   tmp2=sigma*sqrt(T);

#pragma omp for reduction(+:sumx,+:sumx2)
for ( i = 0; i< nSimulations; i++)
{
   double Svals, Pvals;
   Svals = S* exp (tmp1 +tmp2*gaussrand());
   Pvals = Max( Svals - E, 0);
   sumx   +=   Pvals;
   sumx2 += (Pvals*Pvals);
}
stdev = sqrt(((double)nSimulations*sumx2-sumx*sumx)/
   ((double)nSimulations*(double)(nSimulations-1)));

*width = exp(-r *T) * 1.96 * stdev /
                        sqrt((double)nSimulations);
*Pmean = (exp(-r *T) * (sumx / nSimulations));
}
```

The above code calls the function *gaussrand* from within a parallel region and as a consequence the programmer must tell the compiler, through use of the *OpenMP SIMD* extensions, to produce an instance suitable for calling from within the parallel construct:

```
#pragma omp private_function gaussrand private(V1,V2,S)
#pragma omp private_return gaussrand
double gaussrand(void) {
   static double V1, V2, S; static int phase = 0;
   double X;
   ...
   return X;
}
```

Here the application of clause *private_function* expresses the requirement that *gaussrand* is intended to be called from within a parallel region or loop and *private_return* implies that the function will produce a private, i.e. poly, result. This highlights an inportant difference between the threaded approach of *OpenMP*'s out-of-step execution when compared with *OpenMP SIMD*'s lock-step execution.

Using our prototype compiler early benchmarking of the Monte-Carlo simulation for European Option pricing has shown excellent results for the OpenMP SIMD implementation running on ClearSpeed's Advance Board. These benchmarks were performed on a 2.33GHz dual-core Intel Xeon (Woodcrest, HP DL

Table 1. Monte-Carlo - ClearSpeed Board vs. host-only performance comparison

GNU C Compiler with Math Kernel Library	Intel C Compiler with Math Kernel Library	ClearSpeed Advance Board	ClearSpeed Advance Board plus Intel Math Kernel Library
36.6M	40.5M	206.5M	240M

380 G5 system) and the results are shown in Table 1. The results show a 5.15x performance improvement, for 100,000 Monte-Carlo simulations performed using a single ClearSpeed Advance board when compared to simply using an Intel Xeon processor, the Intel C compiler and Math Kernel Library (MKL). When compared against the GNU C compiler and the Intel MKL the performance improvement increases to 5.7x. However, combining both the general purpose Intel processor and MKL with the Advance board achieves the best result at 240M samples per second and a performance factor of up to 6x compared to using just the host processor, Intel C compiler and MKL.

5   Conclusion

In this paper, we have introduced the programming extensions $OpenMP\ SIMD$ for programming arbitrary-sized SIMD arrays. This language is understood by providing an alternative semantics for $OpenMP$. While there have been a number of extensions to existing programming lanuages for accessing the features of SIMD machines, including C^n and C^* [4,13], to our knowledge we are the first to propose using $OpenMP$ to program SIMD machines.

We have developed a set of prototype tools targeting the CSX array architecture and are evaluating using $OpenMP\ SIMD$ as an alternative to the explicit parallel programming language C^n. Early results indicate that it provides at least the same functionality and ability for performance.

$OpenMP$ has a relaxed-consistency shared-memory model that allows threads to have their own view of global data which only required to be synched up at specific points including the explicit flush operation. Any compiler implementing OpenMP SIMD has the ability to promote shared variables to registers and even cache them in PE memory if needed. The flexibility this model allows provides a compiler with the chance to optimize through data migration between mono memory, PE memory and host memory.

In conclusion we believe that there is an important and fascinating concept at the heart of $OpenMP\ SIMD$. Essentially we are translating runtime parallelism into data parallelism. The model can only work if this fundamental difference in the model is not observable. We believe that this is sound as long as we can only observe a program's behavior by its effect on variables. Since we believe we have correctly defined the data parallelism then there should be no observable difference in behavior.

Acknowledgments

We would like to thank Ray McConnell who originally introduced us to OpenMP and David Lacey for his useful feedback on the translation.

References

1. ClearSpeed Technology Plc: White paper: CSX Processor Architecture (2004)
2. Buck, I.: Brook: A Streaming Programming Language. Stanford University (2001)
3. Frigo, M.: Portable High-Performance Programs. PhD thesis, MIT Department of Electrical Engineering and Computer Science (1999)
4. Gaster, B.R., Hickey, N., Stuttard, D.: The C^n Language Specification. ClearSpeed Technology Plc. (2006)
5. OpenMP Architecture Review Board: OpenMP Application Program Interface (2005)
6. Charles, P., Grothoff, C., Saraswat, V., Donawa, C., Kielstra, A., Ebcioglu, K., von Praun, C., Sarkar, V.: X10: an object-oriented approach to non-uniform cluster computing. In: OOPSLA 2005. Proceedings of the 20th annual ACM SIGPLAN conference on Object oriented programming, systems, languages, and applications, pp. 519–538. ACM Press, New York (2005)
7. International Standards Organization: ISO/IEC 98899 - Programming Language C. (1999)
8. Gosling, J., Joy, B., Steele, G., Bracha, G.: The Java Specification, 3rd edn. Addison-Wesley, Reading (2005)
9. Ramanathan, R.: Intel white paper: Extending the world's most popular processor architecture. Intel Corp. (2006)
10. Allen, R., Kennedy, K.: Optimizing Compilers for Modern Architectures. Morgan Kaufmann, San Francisco (2002)
11. Stroustrup, B.: The C++ Programming Language, 2nd edn. Addison-Wesley, Reading (2000)
12. Wilmott, P.: Quantitative Finance, 2nd edn. John Wiley & Sons Ltd., Chichester (2000)
13. Rose, J., Steele, G.: C^*: An extended C language for data parallel programming. In: Internationl Conference on Supercomputing, vol. 2, pp. 72–76 (1987)

OpenMP Extensions for Irregular Parallel Applications on Clusters

Jue Wang, Changjun Hu, Jilin Zhang, and Jianjiang Li

School of Information Engineering, University of Science and Technology Beijing
No.30 Xueyuan Road, Haidian District, Beijing, P.R. China
juewang@acm.org, huchangjun@ies.ustb.edu.cn, zhangjilin@acm.org,
jianjiangli@gmail.com

Abstract. Many researchers have focused on developing the techniques for the situation where data arrays are indexed through indirection arrays. However, these techniques may be ineffective for nonlinear indexing. In this paper, we propose extensions to OpenMP directives, aiming at efficient irregular OpenMP codes including nonlinear indexing to be executed in parallel. Furthermore, some optimization techniques for irregular computing are presented. These techniques include generation of communication sets and SPMD code, communication scheduling strategy, and low overhead locality transformation scheme. Finally, experimental results are presented to validate our extensions and optimization techniques.

1 Introduction

Sparse and unstructured computations are widely used in scientific and engineering applications. This means that the data arrays are indexed either through the value in other arrays, which are called indirection array, or through non-affine subscripts. Indirect/nonlinear indexing causes the data access pattern to be highly irregular. Such a problem is called irregular problem. Many researchers [1] [2] have focused on developing the techniques for the situation of indirection arrays. However, these techniques may not be effective for nonlinear indexing.

OpenMP [3] has been widely recognized as standards in both industry and academe area. For the regular parallel application, a number of efforts [4] [5] have attempted to provide OpenMP extensions on clusters by using SDSM (Software Distributed Shared memory) environment. Using the above methods, the communication overhead among threads may be considerable due to irregular data access pattern. Following their work [6], Basumallik et al. [7] employed producer-consumer flow graph to translate OpenMP programs to MPI. However, this technique requires a private copy in each thread. For the situation where data arrays are indexed through indirection arrays, M. Guo et al. proposed effective OpenMP extensions [8] and corresponding strategy for automatic parallelization [9]. These extensions and strategy can't deduce certain characteristics (e.g., monotonicity) for nonlinear indexing. In this paper, we focus on the OpenMP extensions and parallelism strategy for the applications involving nonlinear indexing.

B. Chapman et al. (Eds.): IWOMP 2007, LNCS 4935, pp. 101–111, 2008.

If the array subscript expressions are nonlinear form, which appear in some irregular parallel applications, the performance of total execution may not be improved using the techniques mentioned above. The following program in Fig. 1 shows a perfectly nested loop Ψ.

/*a perfectly nested loop Ψ */
$$\text{DO } i_1 = 1, N, S_1$$
$$\quad \text{DO } i_2 = L_2(i_1), U_2(i_1), S_2(i_1)$$

$$\cdots\cdots$$

$$\qquad \text{DO } i_n = L_n(i_1, \ldots, i_{n-1}), U_n(i_1, \ldots, i_{n-1}), S_n(i_1, \ldots, i_{n-1})$$
$$\qquad\quad A(f(i_1 \cdots i_n)) = F(B(g(i_1 \cdots i_n)));$$

ENDDO

$$\cdots\cdots$$

ENDDO

ENDDO

Fig. 1. Perfectly nested loop

For the sake of simplicity, we assume that the data array A and B have only one dimension. In the loop, the array access functions (f and g), the lower and upper bound (L, U), and stride (S) may be arbitrary symbolic expressions made up of loop-invariant variables and loop indices of enclosing loops. General parallel compiling techniques can not be applied to these kinds of irregular applications, because there is no affine relationship between the array global addresses of LHS (Left Hand Side) and RHS (Right Hand Side). Commonly, the technology of runtime support library which adopts Inspector/Executor model is applied for communication generation of irregular parallel applications [1][2]. However, the existing libraries have an obvious drawback that significant overhead is caused by the inspector phase.

In comparison to the previous researches, the main characteristics of our work are as follows:

- To the best of my knowledge, we first propose some extensions to OpenMP, aiming at efficient irregular OpenMP codes involving nonlinear indexing.
- In our method, most of the communication set generation and the SPMD code generation can be completed at compile-time.
- A heuristic communication scheduling is used to avoid the communication conflicts and minimize the number of inter-node barrier.
- We introduce a new data reordering scheme for locality optimization.

The above techniques will be presented in the following sections, respectively. These methods can experimentally achieve better performance for the irregular codes involving nonlinear indexing.

The remainder of this paper is organized as follows: Section 2 presents our extensions to OpenMP for irregular application. Section 3 describes the scheme for communication set generation and the SPMD code generation. Section 4 outlines how to reorder data to improve the computation performance. Section 5 introduces the communication scheduling for the performance of inter-node communication. Experimental studies will be shown in Section 6. Finally, Section 7 gives our conclusions.

2 OpenMP Extensions for Irregular Applications

Based on the prior works for multi-paradigm and multi-grain parallelism [10][11], we provide the extensions to OpenMP, the irregular directives, as follows:

$!Eomp parallel do [reduction|ordered] schedule(irregular)

In traditional OpenMP specification, there are four scheduling policies available: static scheduling, dynamic scheduling, guided scheduling, and runtime scheduling. In order to reduce communication overhead and achieve load balance, we add irregular scheduling to OpenMP. This scheduling follows owner-compute rule, where each iteration will be executed by the processor which own the left hand side array reference of the assignment for that iteration. When the parallel region are recognized as an irregular loop at compile-time phase, the compiler will complete the computation for generating communication sets, communication scheduling scheme, and corresponding SPMD code generation at compile-time as much as possible.

```
$! Eomp parallel do order schedule (irregular)
    DO mrs=0, (N*N+N)/2-1
        DO mi=0, N-1
            DO mj=0, mi-1
                mrsij= (mi*mi+mrs*(N*N+N))/2+mj+1
$! Eomp ordered
                xrsij(mrsij) = xij(mj)
$! Eomp end ordered
            END DO
        END DO
    END DO
$! Eomp parallel do
```

Fig. 2. Example 1

Fig. 2 shows a simplified version of loop nest OLDA (from the code TRFD, appearing in Perfect benchmark [12]) using EOMP directives.In this example, the compiler will treat the loop as a partial ordered, i.e. some iterations are executed in ordered way while some other may be executed in parallel. Engelen et al. [13] developed an unified framework for nonlinear dependence testing and symbolic analysis. This technique can be used to derive iteration sets of sequential execution and parallel execution, respectively.

3   Generating Communication Sets and SPMD Code

In the prior works [14], C. Hu et al. proposed a hybrid approach, which integrates the benefits from the algebra method and the integer lattice method, to derive an algebraic solution of communication set enumeration at compile time. They obtain the solution in terms of a set of equalities and inequalities which are derived from the restrictions of data declarations, data and iteration distribution, loop bounds, and array subscript expressions. And based on the integer lattice, they also develop the method for global-to-local and local-to-global index translations. The corresponding SPMD code focuses on two-layer nested loop and can be

```
1.  /*the SPMD code in processor p ( p ∈ {0..n_y − 1}, there exist  n_y processors) */
2.  /*pack and send messages to other processors*/
3.  for  i ∈ {0..n_y − 1}  do
4.          select q using our communication scheduling algorithm in section 4
5.          calculate  Comm_Set(i_1,i_2,...i_x)_[q,p]
6.          for (i_1,i_2,...i_x) ∈ Comm_Set(i_1,i_2,...i_x)_[q,p]
7.                  Append  Comm_Set(B:r − c)_[q,p]  to  buf_send(q)
8.                  Send  buf_send(q)  to processor if not empty
9.          end
10. enddo
11. /*perform local iteration*/
12. calculate  Comm_Set(i_1,i_2,...i_x)_[p,p]
13. In order to improve the locality of local computation, reorder  Comm_Set(A:r − c)_[p,p] to
        generate Comm_Set'(A:r − c)_[p,p] using our locality transformation scheme in section 5.
        According to Comm_Set'(A:r − c)_[p,p], derive Comm_Set'(B:r − c)_[p,p].
14. for (i_1,i_2,...i_x) ∈ Comm_Set(i_1,i_2,...i_x)_[p,p]
15.         Comm_Set'(A:r − c)_[p,p] = Comm_Set'(B:r − c)_[p,p]
16. end
17. /*receive data and perform remaining iterations*/
18. while expecting messages
19.         s = 0
20.         wait for a message from q and store into  Buf_[q](s)
21.         calculate  Comm_Set(i_1,i_2,...i_x)_[p,q]
22.         reorder  Comm_Set(A:r − c)_[p,q] to generate  Comm_Set'(A:r − c)_[p,q]    using our
                locality    transformation    scheme    in    section    5.    According    to
                Comm_Set'(A:r − c)_[p,q], derive  Buf_[q]'(s).
23.         for (i_1,i_2,...i_x) ∈ Comm_Set(i_1,i_2,...i_x)_[p,q]
24.                 Comm_Set'(A:r − c)_[p,q] = Buf_[q]'(s)
25.         end
26.         s = s+1
27. end
```

Fig. 3. SPMD code corresponding to the loop in Fig.1

completed at compile time without inspector phase of run time. In this section, we extend this SPMD code to exploit the parallelism for multi-layer nested loop.

The sending of messages can be performed individually for each remote processor as soon as each packing is completed, instead of waiting for all message for all processors to be packed. The nonlocal iterations can also be performed based on each receiving message, instead of waiting for all messages to be received, because the nonlocal iterations are split into groups based on the sending processors. This strategy can efficiently overlap computation and communication. The SPMD code for the loops in Fig. 1 can be constructed as shown in Fig. 3.

From this SPMD code, $Comm\_Set(i_1,i_2,...i_n)_{[p,q]}$ represents the set of loop indices $(i_1,i_2,...i_n)$that will be executed in the processor p, and that will reference array B elements in the processor q. $Comm\_Set(B : r\text{-}c)_{[p,q]}$ is the set of local array (array B) indices in r-c grid[15], where the row number r specifies the r th round of the elements distributed to a processor and the column number c specifies an element in each round. Similarly, $Comm\_Set(A : r\text{-}c)_{[p,q]}$ is the set of local array (array A) indices in r-c grid. The sequence of message packing $Comm\_Set(B : r\text{-}c)_{[p,q]}$ in the sending processor q is generated by $Comm\_Set(i_1,i_2,...i_n)_{[p,q]}$. In the receiving processor p, the local array A elements with indices $Comm\_Set(A : r\text{-}c)_{[p,q]}$ will access the remote data $Comm\_Set(B : r\text{-}c)_{[p,q]}$ from processor q. The sequence of nonlocal references to array B in the message is also generated by $Comm\_Set(i_1,i_2,...i_n)_{[p,q]}$. Since both sequences are generated by the same $Comm\_Set(i_1,i_2,...i_n)_{[p,q]}$, the receiving processor will access the message in the same order as it is packed in the sending processor according to the same $Comm\_Set(i_1,i_2,...i_n)_{[p,q]}$.

From the above observation, we can learn that the essence of communication pattern in our SPMD code is many-to-many communication, where the numbers of sending and receiving processors are different. This reason is that the real communication doesn't occur among processors when some data size is set to zero. In next section, we will give our communication scheduling algorithm for many-to-many communication.

4   Communication Scheduling

If there is no communication scheduling in many-to-many communication, a larger communication idle time may occur due to node contention and the difference among message lengths during one particular communication step. We revise the greedy scheduling [16] to avoid node contention and reduce the number of barrier and the difference of message lengths in each communication step.

In this paper, we use a communication (COM) table to represent the communication pattern where $COM(i,j)=u$ expresses a sending processor P_i send an u size message to a receiving processor Q_j. According to the COM table, our algorithm generates another table named communication scheduling (CS) table

Algorithm 1. Communication scheduling algorithm for many-to-many communication
   Input: *COM* table
   Output: *CS* table
Begin_of_Algorithm
   Sort messages based on their size in *COM* table and put them into a list *LIST*
   $k = 1$;
   while (there exist unscheduled message) do
        if (the largest message size < the threshold) then
             Put all message in one phase $CS(\cdot ,k)$
        for each unscheduled message in the sorted list *LIST*
             if no conflict, put receiving processor j of the message into $CS(j,k)$
        k++
   enddo
End_of_Algorithm

Fig. 4. Our communication scheduling algorithm

which expresses the message scheduling result. $CS\ (i,k)=j$ means that a sending processor P_i sends message to a receiving processor Q_j at a communication step k. $CS\ (\bullet,j)$ indicate the jth column of CS table. Our scheduling algorithm is shown in Fig.4. If the sizes of the remaining messages are less than a threshold value, all messages are put in one phase.

Similar issue was considered by proposals in [16] to eliminate communication conflicts. The key difference between our algorithm and other scheduling algorithms is that the sending processor is permitted to send messages to different receiving processors in each step when there is no communication conflict. Our algorithm reduces idle time and the number of barrier as much as possible. A motivating example will be given in Fig. 5.

COM	0	1	2	3	4
0	0	3	0	7	3
1	7	0	2	1	2
2	2	2	0	3	7
3	6	0	2	0	0
4	0	4	3	3	0

(a) *COM* table

CS	1	2	3	4
0	3	1	4	
1	0	4	2	3
2	4	3	0	1
3	2	0		
4	1	2	3	

(b) *CS* table derived from
greedy algorithm

CS	1	2	3
0	3	1,4	
1	0	4,2,3	
2	4	3,1	0
3	2	0	
4	1,2		3

(c) *CS* table derived from
our algorithm

Fig. 5. Our communication scheduling algorithm

The COM table in Fig. 5(a) describes a communication pattern. Fig. 5 (b) and (c) present the CS tables derived from greedy algorithm and our algorithm, respectively. In Fig. 6, we show the actual communication scheduling of the

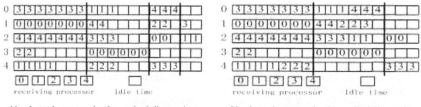

(a) Actual communication scheduling using greedy algorithm

(b) Actual communication scheduling using our algorithm

Fig. 6. Actual communication scheduling

greedy algorithm and our algorithm. The numbered box represents one-unit message and the number in box is the target processor number. The horizontal bar consisting of several consecutive boxes denotes that a sending processor sends a message with different sizes (different number of blocks) to the receiving processors. A barrier (the vertical lines in Fig. 6) is placed between steps. There exists idle time (blank boxes) in each step.

From Fig. 6, we can learn that our strategy not only avoids inter-processor contention, but it also reduces idle time and the number of barrier.

5   Low Overhead Locality Transformation Scheme

In this section, we will present our transformation scheme which relies on deducing the monotonicity of irregular accesses at compile time. Base on the analysis of communication set generation in section 3, we have $Comm\_Set(A)_{[p,q]} = \{f(i_1, i_2, \dots i_n) | (i_1, i_2, \dots i_n) \in Comm\_Set(i_1, i_2, \dots i_n)_{[p,q]}\}$, where $Comm\_Set(i_1, i_2, \dots i_n)_{[p,q]}$ is defined in section 3. According to the monotonicity of irregular accesses, $Comm\_Set(A)_{[p,q]}$ can be divided into n monotonic intervals $(R_1, R_2, \dots R_n)$ acted on by basic operations like union, intersection, etc. Without loss of generality, we can assume that m intervals $(R_1, R_2, \dots R_m)$ are monotonically increasing, and n-m intervals $(R_{m+1}, R_{m+2}, \dots R_n)$ are monotonically decreasing. Fig. 7 presents our data reordering algorithm used in SPMD code (shown in Fig. 3).

The main overhea d of the above algorithm involves the cost of merge-sort algorithm. Our experience tell us that this situation seldom occurs because the rank of nonlinear indexing function (such as f, g in Fig. 1) is often lower than 3. Thus, there is only a little overhead for our locality transform scheme, and it can be overlapped by communication of non-local data.

6   Evaluation and Experimental Results

Based on the prior work [12], we have extended the runtime library routines to support the irregular applications. Our compiler consists of preprocessor, front end, intermediate language, a runtime library, and back end. It uses the approach

Algorithm 2.data reordering algorithm

Input: $R_1, R_2, ..., R_n$, $Comm\_Set(A:r-c)_{[p,q]}$

Output: $Comm\_Set'(A)_{[p,q]}$ which is monotonically increasing

$Comm\_Set'(A:r-c)_{[p,q]}$ corresponding to $Comm\_Set'(A)_{[p,q]}$

Begin_of_Algorithm

/*phase 1 : merge monotone increasing intervals into a monotone increasing interval*/

while (only one interval) do

select two intervals R_i, R_j from monotone increasing intervals

if($up(R_i) < low(R_j)$)

$R_i = R_i \cup R_j$

else if $(low(R_i) > up(R_j))$

$R_j = R_j \cup R_i$

else

/*If the intersection of two intervals R_i, R_j is't null interval, merge these two intervals using

merge sort algorithm*/

$R_i = mergesort(R_i, R_j)$

enddo

/*phase 2: merge monotone decreasing intervals into a monotone decreasing interval*/

while (only one interval) do

select two intervals R_i, R_j from monotone decreasing intervals

if($low(R_i) > up(R_j)$)

$R_i = R_i \cup R_j$

else if $(up(R_i) < low(R_j))$

$R_j = R_j \cup R_i$

else

/*If the intersection of two intervals R_i, R_j is't null interval, merge these two intervals using

merge sort algorithm*/

$R_i = mergesort(R_i, R_j)$

enddo

/*phase 3*/

merge the above two monotone intervals (derived from phase 1 and phase 2, respectively)

into $Comm\_Set'(A)_{[p,q]}$ using merge sort algorithm

/*phase 4*/

According   to $Comm\_Set'(A)_{[p,q]}$,   reordering $Comm\_Set(A:r-c)_{[p,q]}$ to   derive

$Comm\_Set'(A:r-c)_{[p,q]}$

End_of_Algorithm

Fig. 7. Data reordering algorithm

of source-to-source translation and runtime support to implement our extended OpenMP. "whirl" is introduced as the intermediate language from ORC[13] for the optimization work. The runtime library is implemented using our pthread-based library of MPI calls. We evaluated our extensions on the platform, an Ethernet switched cluster with 16 Intel Xeon 3.0G/1024K Cache. Each node has 2GB memory and runs RedHat Linux version FC 3, with kernel 2.6.9. The nodes are interconnected with 1G Ethernets. Time is measured with the *gettimeofday()* system call.

We select code excerpt from TRFD code and a subroutine OLDA (Example 1 shown in Fig.2) from the Perfect benchmarks. In order to evaluate our extensions, we compare the performance of our implementation with that of inspector/executor runtime library (no optimization). Fig.8 shows the average execution time on the cluster for these two implementations.

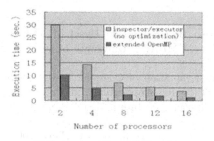

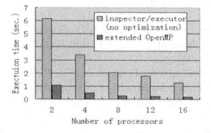

(a) Code excerpt from TRFD code

(b) Subroutine OLDA from the Perfect benchmark (Example 1 shown in Fig.2)

Fig. 8. Comparing average execution time with inspector/executor runtime library (no optimization) and our extended OpenMP

From Fig.8, we observed that our extended OpenMP improves the performance using the optimization techniques in this paper. This phenomenon is more significant when the time steps are larger. Most of computation for Example 1 can't be overlapped by corresponding communication because it takes much more time than communication. Thus, the difference (between inspector/executor and our extended OpenMP) for Example 1 is less than one for subroutine OLDA.

7 Conclusions

The performance of irregular parallel applications (including nonlinear indexing) in current OpenMP implementation has not well investigated. For these irregular codes, current OpenMP only sequentially execute loops by using ordered directives. In this paper, we proposed to extend OpenMP directives to be suitable for compiling and executing irregular codes. Furthermore, we presented some optimization techniques including generation of communication sets and SPMD code, communication scheduling strategy, and low overhead locality transformation scheme. We have introduced these techniques into our extended OpenMP compiler. The experiments validated the efficacy of our extensions and optimization strategies.

Acknowledgments

The work reported in this paper was supported in part by the Key Technologies Research and Development Program of China under Grant No.2006038027015, the Hi-Tech Research and Development Program (863) of China under Grant No. 2006AA01Z105, Natural Science Foundation of China under Grant No.60373008, and by the Key Project of Chinese Ministry of Education under Grant No. 106019.

References

1. Saltz, J., Ponnusamy, R.D., Sharma, S., Moon, B., Hwang, Y.S., Uysal, M., Das, R.: A Manual for the CHAOS Runtime Library, UMI-ACS, University of Manyland (1994)
2. Chakrabarti, D.R., Banerjee, P., Lain, A.: Evaluation of Compiler and Runtime Library Approaches for Supporting Parallel Regular Applications. In: Proc. of the 12th International Parallel Processing Symposium on International Parallel Processing Symposium, pp. 74–80 (1998)
3. OpenMP application program interface, ver 2.5, Tech. report (May 2005), http://www.openmp.org/
4. Chapman, B., Bregier, F., Patil, A., Prabhakar, A.: Achieving Performance under OpenMP on ccNUMA and Software Distributed Shared Memory Systems. Special Issue of Concurrency Practice and Experience, 713–739 (2002)
5. Min, S.J., Basumallik, A., Eigenmann, R.: Optimizing OpenMP Programs on Software Distributed Shared Memory Systems. Int. J. Paral. Prog. 31(3), 225–249 (2003)
6. Basumallik, A., Eigenmann, R.: Towards Automatic Translation of OpenMP to MPI. In: Proc. of the 19th ACM Int'l Conference on Supercomputing (ICS), Boston, pp. 189–198 (2005)
7. Basumallik, A., Eigenmann, R.: Optimizing Irregular Shared-Memory Applications for Distributed-Memory Systems. In: Proc. of the ACM SIGPLAN Symposium on Principles and Practice of Parallel Programming (PPoPP), New York, pp. 119–128 (2006)
8. Guo, M., Cao, J., Chang, W., Li, L., Liu, C.: Effective OpenMP Extensions for Irregular Applications on Cluster Environments. In: Li, M., Sun, X.-H., Deng, Q.-n., Ni, J. (eds.) GCC 2003. LNCS, vol. 3033, pp. 97–104. Springer, Heidelberg (2004)
9. Guo, M.: Automatic Parallelization and Optimization for Irregular Scientific Applications. In: Proc. of the 18th International Parallel and Distributed Processing Symposium (2004)
10. Wang, J., Hu, C., Lai, J., Zhao, Y., Zhang, S.: Multi-paradigm and Multi-grain Parallel Model Based on SMP-Cluster. In: Proc. of IEEE 2006 John Vincent Atanasoff International Symposium on Modern Computing. IEEE Society Press, Los Alamitos (2006)
11. Yongjian, C., Jianjiang, L., Shengyuan, W., Dingxing, W.: ORC-OpenMP: An OpenMP compiler based on ORC. In: Voss, M. (ed.) Proc. Of the International Conference on Computational Science, pp. 414–423. Springer, Heidelberg (2004)
12. Berry, M., Chen, D., Koss, P., Kuck, D., Lo, S., Pang, Y., Roloff, R., Sameh, A., Clementi, E., Chin, S., Schneider, D., Fox, G., Messina, P., Walker, D., Hsiung, C., Schwarzmeier, J., Lue, K., Orzag, S., Seidl, F., Johnson, O., Swanson, G., Goodrum, R., Martin, J.: The PERFECT club benchmarks: effective performance evaluation of supercomputers. International Journal of Supercomputing Applications 3(3), 5–40 (1989)

13. Engelen, R., Birch, J., Shou, Y., Walsh, B., Gallivan, K.: A Unified Framework for Nonlinear Dependence Testing and Symbolic Analysis. In: Proc. of the ACM International Conference on Supercomputing, pp. 106–115 (2004)
14. Hu, C., Li, J., Wang, J., Li, Y.H., Ding, L., Li, J.J.: Communicate generation for irregular parallel applications. In: Proc. IEEE International Symposium on Parallel Computing in Electrical Engineering, Bialystok, Poland, IEEE Society Press, Los Alamitos (2006)
15. Tseng, E.H.-Y., Gaudlot, J.-L.: Communication generation for aligned and cyclic(k) distributions using integer lattice. IEEE Transactions on Parallel and Distributed Systems 10(2), 136–146 (1999)
16. Faraj, A., Yuan, X., Patarasuk, P.: A Message scheduling scheme for All-to-all personalized communication on Ethernet switched cluster. IEEE Trans. Parallel Distrib. Systems (2006)

Optimization Strategies Using Hybrid MPI+OpenMP Parallelization for Large-Scale Data Visualization on Earth Simulator

Li Chen[1] and Issei Fujishiro[2]

[1] Institute of Computer Graphics and Computer Aided Design, School of Software,
Tsinghua University, Beijing 100084, P.R. China
chenlee@mail.tsinghua.edu.cn
[2] Institute of Fluid Science, Tohoku University, 2-1-1 Katahira, Aoba-Ku,
Sendai 980-8577, Japan
fuji@vis.ifs.tohokua.ac.jp

Abstract. An efficient parallel visualization library has been developed for the Earth Simulator. Due to its SMP cluster architecture, a three-level hybrid parallel programming model, including message passing for inter-SMP node communication, loop directives by OpenMP for intra-SMP node parallelization and vectorization for each processing element (PE) was adopted. In order to get good speedup performance with OpenMP parallelization, many strategies are used in implementing the visualization modules such as thread parallelization by OpenMP considering seed point distributions and flow features for parallel streamline generation, multi-coloring reordering to avoid data race of shared variables, some kinds of coherence removal, and hybrid image-space and object-space parallel for volume rendering. Experimental results on the Earth Simulator demonstrate the feasibility and effectiveness of our methods.

1 Introduction

In March 2002, Japan built a supercomputer called the Earth Simulator (ES)[1] for predicting various earth phenomena by numerical simulations. The ES is helpful for solving global environmental problems and taking measures against natural disasters. It has shared memory symmetric multiprocessor (SMP) cluster architecture, and consists of 640 SMP nodes connected by a high-speed network. Each node contains eight vector processors with a peak performance of 8 GFLOPS and a high-speed memory of 2 GB for each processor. According to the Linpack benchmark test, the ES was the fastest supercomputer in the world in June 2002, having a peak performance of 35.61 TFLOPS [1].

ES generates extremely large amount of data everyday. In order to help researchers understand large 3D datasets arising from numerical simulations on the ES, a high-performance parallel visualization library for large unstructured datasets is very necessary. The present study has been conducted as a part of the Earth Simulator project with the goal of developing a parallel visualization library for solid earth simulation.

B. Chapman et al. (Eds.): IWOMP 2007, LNCS 4935, pp. 112–124, 2008.

Although the ES is fast on scientific computing, its graphics processing ability is rather limited. There is no graphics hardware on the ES. Nowadays, more and more attention is focused on the hardware acceleration to achieve real-time speed. However, how to get high speed on this kind of supercomputers without any graphics hardware is still an important topic. It is very critical to make full use of the hardware features of the ES during the implementation of parallel visualization modules.

Different from most supercomputers, the vector processor, 500 MHz SX-6 NEC processor, was adopted on the ES, which contains an 8-way replicated vector pipe capable of issuing a MADD each cycle, for a peak performance of 8 Gflops/s per CPU [2]. The processors contain 72 vector registers, each holding 256 64-bit words. Since the vector performance is significantly more powerful than its scalar one, it is very important to achieve high vector operation ratios, either via compiler discovery or explicitly through code re-organization.

For SMP cluster machines, in order to achieve minimal parallelization overhead, a multi-level hybrid programming model [3,4] is often employed for SMP cluster architectures. The aim of this method is to combine coarse-grain and fine-grain parallelism. Coarse-grain parallelism is achieved through domain decomposition by message passing among SMP nodes using a scheme such as the Message Passing Interface (MPI) [5], and fine-grain parallelism is obtained by loop-level parallelism inside each SMP node by compiler-based thread parallelization such as OpenMP [6]. Another often used programming model is the single-level flat MPI model [3,4], in which separate single-threaded MPI processes are executed on each processing element (PE). The advantage of a hybrid programming model over flat MPI is that there is no message-passing overhead in each SMP node. This is achieved by allowing each thread to access data provided by other threads directly by accessing the shared memory instead of using message passing. However, obtaining a high speedup performance for the hybrid parallelization is sometimes difficult, often more difficult than for flat MPI [7], and hence most applications on SMP machines still adopt a flat MPI parallel programming model primarily due to three reasons. First, although OpenMP parallelization can avoid communication overhead, this method involves thread creation and synchronization overheads; Moreover, if the MPI routines are invoked only outside of parallel regions, all communication is done by the master thread while the other threads are idle; Furthermore, one thread is not able to saturate the total inter-node bandwidth that is available for each node. Many examples show that flat MPI is rather better [3,4], although the efficiency depends on hardware performance, features of applications, and problem size [7].

This paper will describe our methods to re-design visualization algorithms for the ES so as to fit the hybrid architecture of the SMP cluster machines and obtain good parallel performance by hybrid parallel programming model of *MPI + OpenMP*. It is organized as follows. First we give a brief introduction to our parallel visualization library on the ES in Sect. 2. Section 3 gives our method to get good parallel performance for parallel streamline generation module. Section 4 describes our optimization methods for parallel volume rendering large unstructured datasets by the hybrid parallelization. Finally conclusions will be given in Sect. 5.

2   Parallel Visualization Library on Earth Simulator

Currently, most existing commercial visualization software systems work well for relatively small data, but often fail for huge datasets due to the lack of parallel performance and the limitations for some special hardware. We have developed a parallel visualization library which can handle extremely large data size on the ES.

The parallel visualization library provides many visualization functions, including parallel surface rendering and volume rendering [8] for scalar field, parallel streamline, parallel particle tracking and parallel Line Integral Convolution (LIC) for vector fields [9] and parallel hyperstreamlines for tensor fields [10]. Many strategies were adopted to improve visualization quality.

In order to fit for different requirements of simulations, three parallel frameworks are provided in our library including a *client/server* style for interactive visualization, a *concurrent-in-memory* style for extremely large data size, and a *concurrent-in-time* style for real time computational steering on the ES. The *concurrent-in-memory* style is particular powerful on the ES. For extremely large simulation data possibly up to terabytes scale on the ES, it is very difficult and time-consuming to transfer such kind of data to the client machines even for simplified visual primitives. Meanwhile, it is impossible to perform visualization on a single processor client machine due to the memory limitation. Therefore, we have developed the *concurrent-in-memory* style so as to perform the concurrent visualization with computation on the ES. Almost all of the visualization, including data filtering, mapping and rendering, is executed on the ES in highly parallel and images or a sequence of images are generated directly, the size of which is independent of the size of the original dataset. This style is available for arbitrarily large data size.

Our parallel visualization library can be applicable for most typical mesh types used in scientific simulations, including regular grids for Finite Difference Method(FDM), particles for Discrete Element Method(DEM), and unstructured grids for Finite Element Method(FEM). The grids can be very complicated in FEM computing, consisting of unstructured, hierarchical and hybrid grids. Effective visualization algorithms were adopted for different grids, respectively.

In order to make full use of the shared memory architecture and reduce synchronization overhead in each SMP node, we re-design all the visualization algorithms so as to get good thread parallel performance using OpenMP in each SMP node.

3   Optimization for Parallel Streamline Generation

3.1   Motivation

Advection based flow visualization method such as streamlines, streamtubes, particles or flow ribbons [11,12] is still one of the most important flow visualization methods because it can provide intuitive and precise local insight, but it is very difficult to obtain high speedup performance on distributed-memory machines. This is mainly because the streamline is usually approximately constructed by successive integration, as shown in the following formula:

$$p_{k+1} = p_k + h \cdot v_k \tag{1}$$

where p_k,p_{k+1} correspond to the current and next positions on a streamline respectively, h specifies the integration step, and v_k is the velocity vector at the position p_k.

On distributed memory platforms, streamline may pass through many PEs which causes very frequent communications among PEs, as shown in Fig. 1. Before visualization, it is very difficult to know how a streamline runs over the whole field, thus load imbalance is a very critical problem in streamline generation. The speedup performance slows down dramatically with the increase of PE number. Therefore, in order to get better performance for parallel streamline visualization, we should reduce PE number to as few as possible.

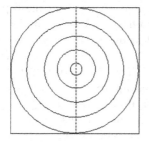 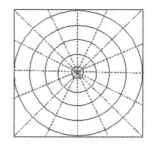

Fig. 1. Different partitions marked by dash lines with the domain number equal to 2 (left) and 16 (right)

The SMP cluster architecture of the ES provides us a good opportunity to reduce the PE number with a ratio of 8. However, because the next position on the streamline is strongly dependent on the current position during the tracing process, how to parallelize with OpenMP in each SMP node is still a problem. We should keep the balanced and independent load for each thread within the SMP node. Meanwhile, the load should also be balanced among SMP nodes. Although all the points' positions on the same streamline is strongly dependent, the positions of points on different streamlines are not related. Therefore, we partition the whole dataset based on the positions of seed points (the starting points for generating streamlines).

3.2 Thread Parallelization Based on Seed Point Distribution

During the streamline generation process, it is very important to decide how many streamlines are generated and where to place the initial points for starting trace so as to reveal the features of the whole field. In our streamline module, the following three styles are provided for determining seed points: (1) User-specified manually: The coordinates of all seed points are specified by users; (2) Semi-automatic: Users specify the number of seed points and the plane equation on which all seed points are located and our module selects optimum positions of all seed points on the specified plane; (3)Automatic: Some existing good seeding methods were implemented in our module to determine seed points automatically according to flow features so as to show the flow more thoroughly.

The seed distribution among PEs is not considered during the process of selecting seed points. They are specified just based on the features of the whole field. Therefore, the even distribution of seeds among processors should be carefully considered in the dynamic repartition process. We implemented a multi-conditional balance algorithm. The priority of evenly seeding is higher than that of vertex number balance in our implementation.

ParMetis [13] was used in our library for dynamic mesh partitioning. It is an MPI-based parallel library that implements a variety of algorithms for partitioning and repartitioning unstructured graphs and dramatically reduces the time cost in communication by computing mesh decompositions such that the numbers of interface elements are minimized. It is an excellent mesh partition tool because it is very fast and supports multi-phase and multi-physics cases. Multiple quantities can be load balanced simultaneously. We assign a relatively larger weight for the vertex very close to or just on seed points. In the multi-constraint graph partitioning of *ParMetis*, a partitioning is obtained such that the edge-cut is minimized and that every subdomain has approximately the same amount of the vertex weights, so this can ensure an almost equal number of seed points located in each subdomain. The number of subdomains is equal to SMP node number and every SMP node is assigned one subdomain.

Inside each domain, thread parallelization by OpenMP is performed during the loop for seed points. For each seed point, a streamline is generated. On the scalar SMP cluster machine, in order to make full use of cache capability, it is better to assign the seed points located closely in positions to the same PE. But the PE of the ES is a vector processor without cache, so data locality is not necessary. Instead, we distribute the seed points with close positions to different PEs. This is because the load on each PE is dependent not only on the number of seed points but also on the length of each streamline inside the PE. If the streamline segment is short within the PE, tracing the seed point is very fast. The streamlines with close distances often have similar flow pattern and thus similar length inside the PE, so scattering close seed points on different PEs sometimes may improve the OpenMP parallel efficiency greatly.

Except seed points, the consideration of flow features is also important for improving parallel performance. For example, around vortex areas, the traced streamline is usually longer than that in straight direction flow. In complicated flow areas, the streamline length inside the current PE is usually longer than that in flat flow areas. Thus, flow feature detection was combined into our flow repartition process. Critical points are found in each processor and larger vertex weights are assigned for the vertices near the critical points. The weight value is inversely proportional to the distance away the critical points. If there are more than one critical point near a vertex, the vertex weight value is the sum of all the computed weights to each neighboring critical point. Using this way, the less vertices can be assigned in the processor which contains complicated flow areas so as to keep the total vertex weights balanced among processors.

3.3   Experimental Results

We have applied the method to a 3D irregular flow field, which simulates a flow passing through an ellipsoidal cylinder. The 3D data set has 20,254,320 unstructured grid cells. Some vortices occur after the flow passes over the cylinder. The seed points are all located upstream at the beginning. Fig. 2 gives the partition examples of 8 domains. We can see the repartitioned mesh by our method can reflect the seed point distribution and flow direction very well. Meanwhile, the vortex areas (located in red and deep blue respectively in Fig. 2(b)) are specially handled to contain less vertices compared with other subdomains.

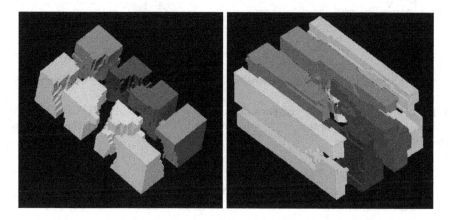

Fig. 2. An example of the flow repartition: original mesh partition without considering seed distributions and flow directions (left); repartitioned mesh by our method (right)

We tested it from 1(8PEs) to 8 SMP nodes (64PEs) on the ES. For generating 800 streamlines using 8 SMP nodes, the optimized hybrid method can achieve 1.54 times faster than the optimized flat MPI method, 2. 98 times faster than the original hybrid method, and 4.11 times faster than the original flat MPI method. The optimized hybrid method is much faster than the original hybrid method mainly because the visualization load is extremely imbalanced in the original mesh partition. For 8 SMP nodes, during almost the whole process, only two nodes are busy and all others are idle if the data is partitioned as Fig.2(a).

Meanwhile, the mesh repartition time is just 1.57 seconds (it still leaves some space for speed improvement) and can be done in preprocessing. From Fig.3, we can see the speedup performance of new repartitioned mesh is much better than original one with the increase of processor number in both flat MPI and hybrid cases. Meanwhile, the optimized hybrid method is much faster than the optimized flat MPI because the communication overhead is still increased very fast with the increase of PE number even if the mesh repartition is used. For the case of using 8 SMP nodes, the whole data is partitioned into 64 domains with flat MPI method and the time cost of communication among 64 PEs becomes large, whereas only 8 domains are used for the hybrid method.

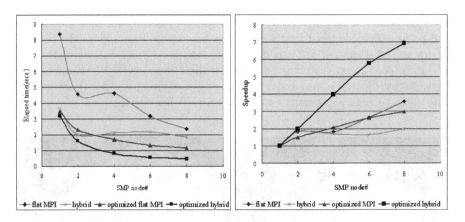

Fig. 3. Parallel performance comparison among flat MPI, hybrid MPI+OpenMP without optimization and optimized hybrid parallel in terms of elapsed time (left) and speedup (right)

4   Optimization for Parallel Volume Rendering Module

4.1   Multi-coloring for Avoiding Data Race

Loop directives by OpenMP [6] are used for the parallelization in each SMP node. A lack of dependency and the absence of a data race are critical in order to achieve efficient parallel performance. However, in our visualization library, data race would exist in some parts. For example, in the PVR module, before ray-casting grids, the gradient value on each vertex must first be computed. The gradient is computed by the shape function of each mesh element. The pseudo algorithm for obtaining gradient values is as follows:

#pragma omp parallel
   for each element i **do**
   begin
      compute the Jacobian matrix of the shape function;

```
for each vertex j of element i do
begin
    for each other vertex k in the element i do
        accumulate gradient value of vertex j contributed by vertex k;
end
end
```

A data race on accesses to the shared variable gradient exists because one vertex is often shared by a number of elements. Although this data race can be avoided by adding mutual exclusion synchronization or writing to a private variable in each thread and then copying to the shared variable, these operations require extra cost with respect to either time or memory. In our parallelization, a multi-coloring strategy was adopted [14], which can easily eliminate data races. As shown in Fig.4, elements are assigned a color such that each element differs in color from its adjacent elements. After coloring, the elements assigned the same color have no common vertex, and so can be parallelized without any data race by OpenMP. Elements in different colors should be computed in serial order. Although extra time is required for multi-coloring, parallel performance can be improved much for large time-step dataset visualizations. The multi-coloring process need only be executed once at the first time-step.

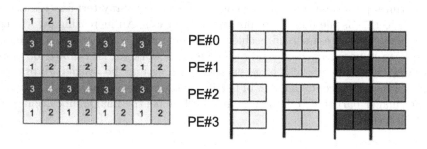

Fig. 4. Multi-coloring for removing data race

4.2   Hybrid-Space Parallel Volume Rendering

Three types of parallelism occur in existing ray-casting PVR algorithms: image-space parallel, object-space parallel and time-space parallel. Image-space parallel subdivides the screen into a number of areas, and assigns one or more areas to each PE, whereas the object-space parallel subdivides all of the volume data into a number of subvolumes, and assigns one or more subvolumes to each PE. Image-space parallel makes load balance easy to obtain and involves less communication, but usually requires replicated data in order to avoid the costly data redistribution. Object-space parallel does not require data redistribution nor replication, so this type of parallelism has better storage scalability as the data size increases. However, since the intensity of each pixel may originate from several PEs, the composition of the final image requires a great deal of communication. The communication overhead will increase greatly with the increase in the PE number.

Obviously, on distributed-memory machines, object-space parallel is better for very large datasets to avoid the memory problem, whereas the image-space parallel is better for shared-memory machines. However, for SMP cluster machines, the hybrid architecture can take advantage of both parallel methods. Object-space parallel can be used among SMP nodes to reduce the data size on each SMP node, and image-space parallel is used in each SMP node to take advantage of its shared memory. By hybrid-space parallel, we can avoid the storage problem for large datasets, the communication bottleneck for large numbers of PEs, and obtain good load balance in each SMP node.

Furthermore, hybrid-space parallel with MPI+OpenMP parallelization can obtain much higher performance due to the early ray termination algorithm [8]. The ray casting volume rendering method is adopted in our PVR module so as to obtain high quality rendering images. Early ray termination is a widely used acceleration technique in the ray casting volume rendering. When we cast a ray and let it pass through the volume data, the current accumulated opacity value will be computed. If the opacity approaches zero, it means the elements behind cannot contribute anything to the final image, so we can stop the tracing directly. This can avoid much useless computation and usually may result in 2-3 times performance improvements. However, early-ray termination becomes less effective as the number of processors increases. An increasing percentage of the processors are assigned occluded portions of the data and therefore either perform unnecessary computation or have no work to do. So 50% or more of the processing power of a large parallel machine may be wasted. Hybrid-space parallel can reduce the number of partitioned domains with a ratio 8. Meanwhile, the smaller partitioned domain number will make the final composition process much faster.

4.3   Revise Ray-Tracing Algorithm for Avoiding Coherence

In order to get high parallel performance in each SMP node, algorithms need be changed in many places for parallel visualization modules. For example, exploring and making full use of many kinds of coherence has been an important focus in the area of computer graphics, whereas dependency removal is important for getting both thread parallel and vector performance. Therefore, different algorithms were adopted in our PVR method for flat MPI and *MPI+OpenMP+vectorization* respectively.

As shown in Fig.5(a), during the ray-casting process in the PVR method, all the sample points along the ray need to be composed from front to back. There are some good methods to quickly decide the next position the ray reaches by the current position and ray direction. This can accelerate the PVR method greatly because the intersection computation spends much time in the ray-casting PVR method.

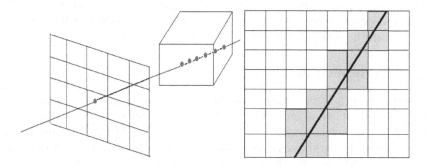

Fig. 5. Different algorithms used for the ES visualization modules: exploring coherence (left) and avoiding dependence on the ES (right)

However, it cannot get good performance on the ES. Because the position of the next sample point is strongly dependent on the previous one, the vectorization and thread parallelization cannot be applied. Therefore, we have to give up the coherence merit and avoid dependency as possible as we could. As shown in Fig.5(b), on the ES, first we determine the set of possible voxels according to the first and last intersection points in the supervoxel. The number of possible voxels can be further reduced by judging whether the distance between the center point of the voxel and the ray line is shorter than the half of the longest diagonal line of the voxel. The possible voxels are marked in dark color in Fig.5(b). Then find intersection points with each possible voxel independently and store the results for all the intersected voxels. Finally make composition from front to back. The

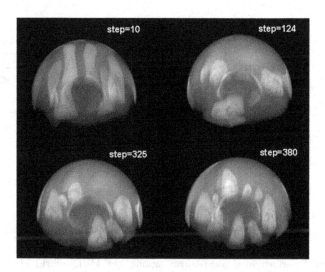

Fig. 6. Volume rendered images for time-varying geo-dynamo simulation by our PVR module

correct composition order can be decided by index values in regular grids. After revision, dependency only exists in the last step. It can achieve better speed on the ES than the original method.

4.4   Experimental Results

We have tested the proposed methods by a time-varying geo-dynamo simulation dataset. This dataset is generated to simulate time-evolution of vorticity for the geo-dynamo in a rotation hemi-spherical shell. It has 18,027,520 unstructured grid elements. The grid size is much different in this dataset, in which the largest one is about 500 times larger than the smallest one.

Volume rendered images for 500 time-steps with 32 rotation frames per time-step are tested using from 1(8PEs) to 24 SMP nodes (192 PEs) on the ES. Fig.6 shows four images at different time-steps. The pattern of vorticity and its distribution on the inner and outer boundary surfaces can be revealed very clearly by these images. To generate almost the same quality 16,000 images up to 24 SMP nodes, the comparison in terms of elapsed time and speedup performance by the flat MPI method, hybrid *MPI+ OpenMP* without optimization and hybrid

Table 1. Elapsed time to generate PVR images for 500 time-steps with 32 rotation frames per time-step by the three methods for the same geo-dynamo simulation using different SMP node numbers

	1 node	2 nodes	4 nodes	8 nodes	12 nodes	16 nodes	20 nodes	24 nodes
Flat MPI (s)	207566	117483	68410	49197	40288	34628	32454	31217
Hybrid (s)	173106	89982	47230	25108	19177	15253	12498	10256
Optimized(s)	159232	80103	42712	23746	17523	14152	11238	9808

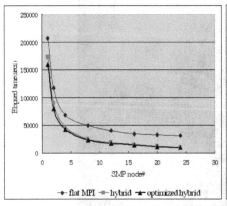 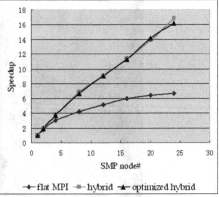

Fig. 7. Parallel performance comparison among flat MPI, hybrid MPI + OpenMP without optimization and optimized hybrid parallel in terms of elapsed time (left) and speedup (right)

MPI+OpenMP with optimization are shown in Fig. 7. The elapsed time of the three methods is listed in Table 1. We can see the hybrid *MPI+OpenMP* parallelization is much faster than flat MPI due to the hybrid-space parallelization and early-ray termination. Moreover, after further optimized by the multi-coloring reorder and coherence removal, the optimized hybrid method achieves better results.

5 Conclusions

This paper describes some strategies for improving speedup performance of hybrid *MPI+OpenMP* parallelization in the implementation of a visualization library for the ES, including the flow field repartition considering seed point distributions and flow features for parallel streamline generation, multi-coloring reorder to avoid data race of shared variables, the coherence removal, and hybrid image-space and object-space parallel for volume rendering. Future work will focus on performance tests on large simulations using more than 100 SMP nodes and improving the vectorization performance for each PE.

Acknowledgement

This study is a part of the "Frontier Simulation Software for Industrial Science" project in the "IT Program" funded by the Ministry of Education, Culture, Sports, Science and Technology, Japan (MEXT). It is also supported in part by the NSFC through grants No. 60773143 and No. 90715043. Furthermore, the authors would like to thank the reviewers of the International Workshop on OpenMP 2007 to give us many valuable comments for rewriting the paper.

References

1. Earth Simulator Research and Development Center Web Site,
   http://www.es.jamstec.go.jp/
2. Oliker, L., Carter, C., Shalf, J., Skinner, D., Ethier, S., et al.: Evaluation of Cache-based Superscalar and Cacheless Vector Architectures for Scientific Computations. Supercomputing, 38–48 (2003)
3. Nakajima, K.: OpenMP/MPI Hybrid vs. Flat MPI on the Earth Simulator: Parallel Iterative Solvers for Finite Element Method. In: International Workshop on OpenMP: Experiences and Implementations (WOMPEI 2003). LNCS, vol. 2858, pp. 486–499. Springer, Heidelberg (2003)
4. Cappelo, F., Etiemble, D.: MPI versus MPI+OpenMP on the IBM SP for the NAS Benchmarks. Supercomputing, 12–19 (2000)
5. MPI Web Site: http://www.mpi.org
6. OpenMP Web Site: http://www.openmp.org
7. Rabenseifner, R.: Communication Bandwidth of Parallel Programming Models on Hybrid Architectures. In: International Workshop on OpenMP: Experiences and Implementations (WOMPEI 2002). LNCS, vol. 2327, pp. 437–448. Springer, Heidelberg (2002)

8. Levoy, M.: Display of Surfaces from Volume Data. IEEE Computer Graphics and Applications 8(3), 29–37 (1988)

9. Cabral, B., Leedom, C.: Image Vector Field Using Line Integral Convolution. In: Computer Graphics Proceedings, ACM SIGGRAPH, pp. 263–272 (1993)

10. Delmarcelle, T., Hesselink, L.: Visualizing Second-Order Tensor Fields with Hyper-Streamlines. IEEE Computer Graphics and Applications 13(4), 25–33 (1993)

11. Schroeder, W.J., Volpe, C.R., Lorensen, W.E.: The Stream Polygon: A Technique for 3d Vector Field Visualization. In: Proc. Visualization, vol. 91, pp. 126–132. IEEE Computer Society Press, Los Alamitos (1991)

12. van Wijk, J.J.: Flow Visualization with Surface Particles. IEEE Computer Graphics and Applications 13(4), 18–24 (1993)

13. http://www-users.cs.umn.edu/~karypis/metis/parmetis/

14. Washio, T., Maruyama, K., Osoda, T., Shimizu, F., Doi, S.: Blocking and Reordering to Achieve Highly Parallel Robust ILU Preconditioners. In: RIKEN Symposium on Linear Algebra and its Applications, The Institute of Physical and Chemical Research, pp. 42–49 (1999)

An Investigation on Testing of Parallelized Code with OpenMP

Robert Barnhart, Christian Trefftz, Paul Jorgensen, and Yonglei Tao

Department of Computer Science and Information Systems
Grand Valley State University, Allendale MI 49401-9403
{trefftzc,jorgensp,taoy}@gvsu.edu

Abstract. Testing is a crucial element of software development. As OpenMP becomes more widely used, a relevant question for developers is: How will programs, that have been parallelized with OpenMP, be tested for correctness? OpenMP programs are concurrent programs and as such all the risks of concurrent programming are present as well. This paper presents some observations about testing parallelized loops using OpenMP.

1 Introduction

OpenMP is a programming environment made up of compiler directives, run-time library routines, and environment variables that allows parallel processing on shared memory computer systems [1]. Because of the concurrent nature of the environment, there are data dependencies [2,3] that, if not handled correctly, can cause faults in a program. Standard testing plans [4], both of the structural and functional type, are not suitable to be used for programs which make use of OpenMP because of the complications introduced by the use of multiple threads processing the same code at the same time [1]. While much research has been done to help a programmer detect, classify, and remove these data dependencies, not much has been done to help test a program that has been parallelized. This paper explores different approaches to testing to see what might be done to discover improperly parallelized loops. The two most widely used testing methods - functional (black box) and structural (white box) testing [4] - might not be sufficient when it comes to testing programs that have been parallelized using OpenMP. It is possible to use a valid set of functional test cases on both a correct serial version and an incorrectly parallelized version of a program. The tests may produce correct results, even though the parallel version is incorrect, because the data dependencies produced by parallelizing a program are not likely to become apparent every single time the program is tested.

Typical scientific and engineering applications spend most of their time in a small set of loops. Parallelizing these loops yields, usually, reasonable speedups

---

[1] There is a commercial product from Intel that checks the correctness of programs that use threads, based on the analysis of memory access from the threads. But the problem remains for other brands of processors.

B. Chapman et al. (Eds.): IWOMP 2007, LNCS 4935, pp. 125–128, 2008.

[3]. Therefore, we will concentrate here on the parallelization of loops with OpenMP.

In the realm of sequential processing, J.C. Huang's rule [5] is applied when testing loops. Huang proved that every loop involves a decision, and in order to test the loop, we need only to test both outcomes of the decision: one outcome is to traverse the loop (only traversing it one time is sufficient), and the other is to exit (or not enter) the loop. When testing a parallelized loop, however, Huang's idea is not applicable. In fact, it might create a false sense of security, as the problems caused by parallelization will *not* surface until *at least* two iterations of the loop are executed concurrently.

Thus if one traversal of a parallelized loop is not good enough, how do we prove that the parallelization of that loop is correct? This is the key question being addressed in the following section.

2   Testing Parallelized Loops

We tried, experimentally, to determine if a particular number of runs of a program can be recommended to testers who want to be sure that a program has been parallelized appropriately. We performed experiments in which we examined the following factors: the computer architecture involved, the number of processors being used, the number of iterations in the loop, and the amount of processing done in the loop.

To perform our tests, we decided to take the following sequence of steps:

- Take different kinds of data dependencies.
- Test each dependency in multiple environments. We had access to two different environments. The first computer used in testing was a laptop, an IBM Thinkpad with a 1.8 GHz processor, 512 MB of RAM, and Windows XP. We also had access to another computer, a multiprocessor with two 2.8 GHz Xeon multithreaded processors, 6 GB of RAM, and Windows XP. Visual Studio 2005 was used on both machines to compile and execute the tests.
- Vary the number of threads used, from two to ten, and all the way to 62 in one test by using a directive from OpenMP.
- Vary the number of iterations in the loop from 10 to 100 million.
- For different combinations of factors above, find out how many tests it takes to uncover the error.
- Run each test 50 times, and take the average. This average is what can be found in the tables included below.

The following four pieces of code were used for the tests:

1. Summation:

```
#pragma omp parallel for
for (int i = 0; i < n; i++)
  x = x + 1;
```

2. Array Shift:

```
#pragma omp parallel for
for(int i = 0; i < (n-1);i++)
  a[i] = a[i+1];
```

3. Summation Plus: This loop is similar to the first segment of code, but it adds a simple inner loop, which effectively adds 100 statements to the work being done in the parallelized loop.
4. Array Shift Plus: This loop is similar to loop 2, with a simple inner loop to increase the amount of work being done within the parallelized loop.

Due to limited space, we show only the results of the first test below. The following tables contain the results of the test. The different rows show how the results changed with the size (number of iterations i.e. n in the test code) of the loop. The columns show how the results changed with the number of threads that were used to process the loop. The actual table cell values show how many times (on average) the loop as a whole had to be executed for the particular combination of number of iterations and threads to result in an error. An error in this set of tests meant that the output from the execution of the parallel program is different from the output from the execution of the sequential version of the code. A tester wants the number of times that the program needs to be executed to be small as the goal is to find the minimum number of times the loop must be executed in order to be certain that the loop is correct.

Table 1. Loop 1 (Summation) Test Results on a laptop

Iterations \ Threads	2	4	6	8	10
100	232391	47617	62195	45124	56529
1,000	40507	2106	1554	855	854
10,000	4351	469	85	77	62
100,000	373	48	8	8	7
1,000,000	21	3	1	1	1
10,000,000	1	1	1	1	1
100,000,000	1	1	1	1	1

Several observations about the test results might be useful to a tester. The number of threads used to execute the loop significantly affects the ability for a tester to uncover a problem when the loop has a small number of iterations. Increasing the number of threads from two to four has a major positive impact on testing; but using any more than six threads does not have any advantages, at least in the environments where the tests were run; in fact, in some situations it even makes things worse. Also, as the number of iterations increases, the advantage of using more threads decreases.

We believe that the previous observations can be used as guidelines. For example, if the tester can control both the number of iterations and the number

Table 2. Loop 1 (Summation) Test Results on a multiprocessor

Iterations \ Threads	2	4	6	8	10
100	1283401	10	1	3	4
1,000	27548	1	1	5	1
10,000	1	1	1	1	1
100,000	1	1	1	1	1
1,000,000	1	1	1	1	1
10,000,000	1	1	1	1	1
100,000,000	1	1	1	1	1

of threads used, the best situation, regardless of the environment, will be to use two to four threads, and to force the loop to iterate a few million times. Based on this set of experiments, this combination would be likely to uncover an improperly parallelized loop. Running that same test multiple times would increase the tester's level of confidence.

Sometimes, however, the tester cannot control all of the variables in the testing environment. This would require a larger number of tests.

The results are, in a sense, discouraging. A developer who wants to test his code to ensure that it is correct, should run his program a large number of times on a multiprocessor. But there is no magic number of runs that will ensure that the parallelization is correct. The absence of errors after a large number of runs does not guarantee that the code is correct. Better, more rigorous and robust approaches are required.

3 Conclusion

Testing code that has been parallelized with OpenMP introduces new challenges to the testing process. Mistakes will not be easy to detect using traditional testing techniques as section two shows. Further research on more rigorous and formal methods to test code that has been parallelized with OpenMP is required. Running parallel code a large number of times alone may not be sufficient nor practical to uncover errors.

References

1. Chandra, R., et al.: Parallel Programming in OpenMP. Morgan Kaufmann, San Francisco (2001)
2. Tomasulo, R.: An Efficient Algorithm for Exploiting Multiple Arithmetic Units. IBM Journal of Research and Development 11(1), 25 (1967)
3. Padua, D.A., Wolfe, M.: Advance Compiler Optimization for Supercomputers. Communications of the ACM 29(12), 1184–1201 (1986)
4. Jorgensen, P.: Software Testing - A Craftsman's Approach, 2nd edn. CRC, Boca Raton (2002)
5. Huang, J.: A New Verification Rule and Its Application. IEEE Transactions on Software Engineering 6(5), 480–484 (1980)

Loading OpenMP to Cell: An Effective Compiler Framework for Heterogeneous Multi-core Chip

Haitao Wei and Junqing Yu*

College of Computer Science & Technology,
Huazhong University of Science & Technology, Wuhan, 430074, China
whtaohust@163.com, yjqing@hust.edu.cn

Abstract. The Cell employs a multi-core-on-chip design that integrates 9 processing elements, local memory and communication hardware into one chip. We present a source to source OpenMP compiler framework which translates the program with OpenMP directives to the necessary codes for PPE and SPE to exploit the parallelism of a sequential program through the different processing elements of the Cell. Some effective mapping strategies are also presented to conduct the thread creating and data handling between the different processors and reduce the overhead of system performance. The experimental results show that such compiler framework and mapping strategy can be effective for the heterogeneous multi-core architecture.

1   Introduction

In recent two years, many processor vendors have pushed out multi-core chips with multiple processors to improve the performance and get more efficient energy. As one kind of heterogeneous architecture, Cell processor integrates one PowerPC Processor Element (PPE) and eight Synergistic Processor Elements (SPE) into one chip. To effectively develop applications on Cell, a high-level programming model is demanded urgently, and OpenMP seems to be a good candidate.

There are a number of source-open OpenMP implementations to translate a C program with OpenMP directives to a C program with Pthreads. In our compiler framework, a program with OpenMP directives is separated to two programs, one is for PPE, and the other one is for SPE. The SPE runtime library customized for the heterogeneous feature of the Cell is utilized to build the OpenMP runtime library.

The rest of this paper is organized as following. Section 2 introduces the Cell architecture and the source to source compiler design. The experimental results are presented in Section 3. Section 4 concludes this paper.

2   The OpenMP Compiler on Cell

Cell processor consists of a 64-bit multithreaded Power PowerPC processor element (PPE) and eight synergistic processor elements (SPE) connected by an

---

* Corresponding author.

B. Chapman et al. (Eds.): IWOMP 2007, LNCS 4935, pp. 129–133, 2008.

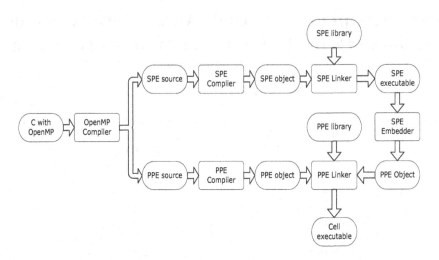

Fig. 1. The OpenMP Compiler Framework for Cell

internal, high-bandwidths Element Interconnect Bus (EIB) [2]. The PPE supports Vector/SIMD Multimedia Extension (VMX) instruction set to accelerate multimedia and vector application [3]. Each SPE contains non-coherent local store for instructions and data. SPE's instruction and data is transferred through DMA from PPE's main storage to SPE's local storage. PPE and SPE execute instructions in parallel.

The OpenMP Compiler Framework on Cell similar to CellSs [4] is illustrated in Fig. 1. Contrasting to CellSs constituted of CellSs compiler and Cell back-end compilers, the general structure of our compiler framework is composed of two components: a source to source OpenMP translator/compiler front-end and PPE-SPE combination compilers. Given a sequential C program with OpenMP directives, the source to source OpenMP translator/compiler is used to generate two program sources. The one is PPE source corresponding to the master thread code, and it is compiled by the PPE compiler to generate a PPE object. The other one is SPE source corresponding to the slave thread code, and it is compiled by the SPE compiler to obtain a SPE object, which can be linked with the SPE library to get a SPE executable code. Finally, the SPE embedder is used to generate a PPE object, which is linked by the PPE linker to generate the final executable code together with other PPE objects and PPE library.

Non-coherent memory and separate code running on Cell architecture, brings a new challenge to OpenMP compiler. The OpenMP compiler must tackle with the respective code generation for PPE and SPE, data distribution, and synchronization.

Parallelization Division. The parallel computation needs to be split among PPE and SPE processors. In Cell platform, PPE and SPE processors can be viewed as a group of "threads". The PPE starts to execute the program first, with spawning a number of SPE spinner threads which sleep until they are needed. When a parallel construct is encountered, the SPE threads are woke up to execute the parallel code.

Data Distribution. As the local storage belongs to the corresponding SPE, and the main storage is shared by PPE and SPE processors, private data is allocated in the stack and static storage of each SPE's local storage. Shared data is allocated in the main storage. As each SPE fetches data for computation from the local storage, a copy of the shared data is allocated in each SPE's local storage by DMA transfer.

Synchronization. The mailbox and event signaling mechanism supported by the hardware of Cell can be utilized to synchronize between SPEs and PPE. When the "fork" point is encountered, the PPE places require in each SPE mailbox entry to spawn a group of SPE threads to execute the corresponding task. As to the "barrier" synchronization, each SPE thread notifies the PPE thread by signals or mailbox.

3   Experiments and Results

In this section, mapping strategies of locality and event mechanisms are proposed to reduce the overhead of shared data and parallel synchronization for OpenMP program running on Cell. The experimental results of performance overhead are measured based on the cycle accurate Cell Simulator 2.0 provided by IBM [5].

Exploitation Locality. The solution is to keep a copy of the shared data in the local storage of each SPE instead of the direct accessing to main storage [1]. If the local copy of the shared data is updated, the changed data will be written through DMA transfers back to the main storage after the end of parallel section.

Fig. 2 illustrates the overhead due to the exclusive accessing to the shared data adopting two different mechanisms: atomic accessing and local accessing with mutex variable and DMA transfer. Loop iteration, in spite of a simple operation, may be used in most programs. In this case, a shared integer array with 128 entries is allocated in the main storage, and each SPE executes the iterations of addition operation on each entry of the shared array. By setting 10 iterations, the overhead of the atomic (ATOMIC1) and local (LOCAL1) mechanism for shared data accessing are measured almost the same. But with the number of iteration increasing to 100, the overhead of the atomic (ATOMIC2) mechanism increases sharply, especially the number of the SPU reaches 6 and 8. Contrast to the atomic mechanism, the overhead of local (LOCAL2) mechanism is almost the same as the 10 iteration one (LOCAL1), except the additional accessing time for more iteration. The local mechanism for shared data accessing experiment reduces 40% overhead for 8 SPEs. It is shown that with the accessing frequency increasing, a larger overhead reduction can be achieved.

Parallel Synchronization. The parallel synchronization is implemented with the mechanisms of memory tag polling and event trigger on Cell. The first method is to allocate a tag memory for each SPE with values initialized to

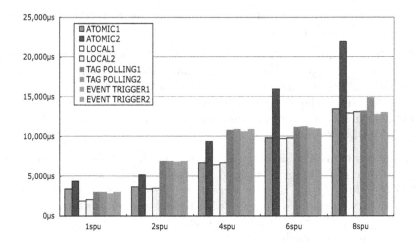

Fig. 2. Overhead of the Shared Data Accessing and the Parallel Synchronization

"0" and each SPE polls the tag to wait the task assigned by PPE. PPE wakes up each SPE by setting the tag to "1". If the tag is detected to be "1", each SPE runs the computation immediately. After the computation is completed, the tag is set to "2" to notice PPE. At the same time, the PPE polls the tag until all of them are set to "2" by the SPEs, and then the PPE cleans the tag to "1" to resume the SPE execution. The second method is to utilize the mailbox and signal mechanism for parallel synchronization. The PPE deploys the task to each SPE by sending signal, once receiving the "start" signal each SPE does the work from PPE. And at the time of barrier, each SPE will send "completion" event to PPE by mailbox and then sleep. Once receiving all the events from SPEs, PPE will resume the SPEs to continue to work.

As Fig. 2 illustrates, the overhead of the parallel synchronization is measured with the mechanism of the memory tag polling and event trigger. The PPE assigns the parallel work of array summing (like Exploitation Locality) to each SPE, and the synchronization operation occurs at the start and the end of parallel work. With the calculation increase from 10 to 100 iterations, the tag polling mechanism (from TAG POLLING1 to TAG POLLING2) and event trigger mechanism shows the similar execution overhead, except the abrupt overhead increase with 8 SPE. But, it is believed that the event trigger mechanism will be a better candidate for parallel synchronization, besides the less little reduction than tag polling, it also saves the storage and execution time because of its un-blocked execution.

4   Conclusions

This paper presents a source to source OpenMP compiler framework which translates the program with OpenMP directives to the necessary codes for PPE and

SPE, respectively for Cell EA. The OpenMP compiler framework proposed is amazing, but there is a lot of work left, as for example: the effective code generation method and the improvement of data distribution and synchronization mechanism between the PPE and the SPEs.

References

1. Eichenberger, A.E., O'Brien, J.K., et al.: Using advanced compiler technology to exploit the performance of the cell broadband engine architecture. IBM System Journal 45(1), 59–84 (2006)
2. Pham, D., et al.: The Design and Implementation of a First-Generation CELL Processor. In: Proceeding of 2005 IEEE International Solid-State Circuits Conference, pp. 184–185 (2005)
3. IBM Corporation: PowerPC Microprocessor Family: AltiVec Technology Programming Environments Manual (2004)
4. Bellens, P., Perez, J.M., Badia, R.M., Labarta, J.: CellSs: a Programming Model for the Cell BE Architecture. In: Proceedings of 2006 ACM/IEEE conference on Supercomputing, pp. 200–210 (2006)
5. IBM Corporation: IBM Full-System Simulator User's Guide version 2.0 (2006)

OpenMP Implementation of Parallel Linear Solver for Reservoir Simulation

Changjun Hu, Jilin Zhang, Jue Wang, and Jianjiang Li

School of Information Engineering, University of Science and Technology Beijing
No.30 Xueyuan Road, Haidian District, Beijing, P.R. China
huchangjun@ies.ustb.edu.cn, zhangjilin.bj@gmail.com

Abstract. In this paper, we discuss an OpenMP implementation of an evolutionary LSOR method, the MBLSOR method, for solution of system of linear equations related to reservoir simulation on SMPs. MBLSOR method not only can improve the data locality by spatial computational domain decomposition technique, but it also can parallel the sub blocks with no data dependence. We compare the performance of different parallel LSOR methods in terms of efficiency and data locality. Numerical results on SMPs indicate that MBLSOR algorithm is more efficient.

1 Introduction

Reservoir simulation is a very compute-intensive application often requiring large amounts of CPU time on state of the art supercomputers. Many researchers have focused on developing the techniques for the parallel reservoir simulation [1]. J. Shu et al. [2] developed two domain decomposition methods on shared-memory parallel computer systems for solving a black numerical simulation of reservoir.

OpenMP [3] is an industry standard for shared memory parallel programming agreed on by a consortium of software and hardware vendors. It is considerably easier for a non-expert programmer to develop a parallel application under OpenMP. OpenMP also permits the incremental development of parallel code. Thus it is not surprising that OpenMP has quickly become widely accepted for shared-memory parallel programming.

In this paper we present a new parallel implementation of LSOR linear system solver in simulator optimized for SMPs using OpenMP. We use an evolutionary technique to parallelize the LSOR solver for linear banded system, which take most of the total execution time of a typical simulation in SMPs. Results show that the parallelization strategy can improve data locality and exploit parallelism.

The remainder of this paper is organized as follows: In Section 2 we present a description of linear solver. The parallelization solver and the comparison with other parallel methods are detailed in Section 3. Section 4 shows simulation results and the corresponding analysis, and Section 5 gives our conclusions and future works.

B. Chapman et al. (Eds.): IWOMP 2007, LNCS 4935, pp. 134–137, 2008.

2   Description of the Linear Solver

The simulator we have studied uses finite difference approximations to convert the black oil simulator partial differential equations to the algebraic equations.

$$AE_iP_{i+1} + AW_iP_{i-1} + AN_jP_{j+1} + AS_jP_{j-1} +$$
$$AB_kP_{k+1} + AT_kP_{k-1} + E_{i,j,k}P_{i,j,k} = b_{i,j,k} \tag{1}$$

The solution of LSOR method in the simulator is oriented implicitly along x-axis and explicitly along other axes. For an x-line (along the x-axis) LSOR the linear matrix system of equation reduces to the following tridiagonal system:

$$AW_{i,j,k}P_{i-1,j,k}^{n+1} + E_{i,j,k}P_{i,j,k}^{n+1} + AE_{i,j,k}P_{i+1,j,k}^{n+1} = b_{i,j,k} - NP_{i,j,k}^n \tag{2}$$

P^{n+1} and P^n represent the pressure matrices at n+1 and n time step respectively. This tridiagonal system is then solved by direct method in simulator. As an extension of SOR theory, LSOR require that all points in a line be solved first implicitly (using a tridiagonal system of equations), and then these values would be appropriately weighted using ω. As we proceed from line to line, the updated values of the unknowns from the previous line could be used.

3   Parallel Implementation of MBLSOR

The LSOR(SOR) method is inherently sequential due to embarrassingly parallel, and data locality is worse in three dimensions when the number of grids increases. Many researchers have focused on developing the techniques for the parallel SOR method[4-7]. The MBLSOR (Multi Block Line Successive Over Relaxation) method, which we have developed, can improve three different performance aspects: inter-iteration data locality, intra-iteration data locality and parallelism. Intra-iteration locality refers to cache locality upon data reuse within a convergence iteration, and inter-iteration locality refers to cache locality upon data reuse between convergence iterations. The MBLSOR method executes convergence spatial computational domain iterations in a sub-block-by-sub-block fashion. Each sub-block contains computations from multiple convergence iteration for a sub set of unknowns along one axis. Using this algorithm, the above three performance aspects can be obtained by selecting the sub-blocks for improved inter-iteration locality and ordering lines of unknowns for intra-iteration locality, and executing the independent sub-block in parallel.

Implementation of this algorithm is straightforward as follows:

Step 1: Divide the spatial computational domain into desired sub-domains. The whole domain A is partitioned into m (the number of CPU) sub-domains. Each sub-domain owns T/m x-lines of unknowns. To be ensured load balance of computation between CPUs, we should amend the partition of domain A. the first CPU has $T/m - (k/m)$ x-lines of unknowns ,the last CPU has $T/m + (k/m)$ x-lines, and others have T/m x-lines.

Step 2: Divide the spatial computational sub-domain to desired sub-blocks. If the size of cache is equal to the memory requirements for b x-lines of unknowns, the number of sub-block will be $T/m(b - k)$.

Step 3: Indicate the number for sub-blocks. The amendment algorithm use the numbering method to indicate the execution order for adjacent sub-block due to data dependence, thus the partition numbering affects the data dependence direction between sub-blocks. In order to exploit parallelism, number of each sub-block must be given by using $cyclic(m)$ method.

Step 4: Amend the boundary of each spatial computational sub-block to reuse data before the cache-line is evicted.

Step 5: Parallel update sub-blocks where there is no data dependence. After the first m sub-blocks parallel executed, other sub-blocks computation in difference sub-domains can be parallel executed but sub-blocks in one sub-domain must be serial Updated.

Step 6: Using step 5, if it is not convergent we can update the next spatial computational domain.

For shared memory parallel processing, the synchronization time is the amount of time that processors are waiting for data dependent results that are generated by other processors. The minimized synchronization time can improve parallel efficiency. For domain decomposition LSOR, there is intra-iteration synchronization because adjacent cells between different partitions depend on each other. For above two traditional parallel methods, the synchronization barriers between all convergence iterations also cause parallel inefficiency. But for MBLSOR method, there are only synchronization issues between sub-domains and between several convergence iterations.

Both multi-color and domain decomposition LSOR methods must execute all the iteration points within one convergence iteration before the next convergence iteration. If the subset of unknowns assigned to a processor does not fit into cache, no inter-iteration locality occurs. MBLSOR method can update multiple convergence iterations over a subset of the unknowns, and the size of sub-block can fit into cache. Thus, theoretically, MBLSOR can improve the inter-iteration locality. In next section, we will validate the above analysis.

4   Performance Results

In this section, we show simulation results for the MBLSOR, which used in the reservoir simulator with OpenMP directives. All performance tests were conducted on a SMP system with Intel Xeon 3.0G/512K L2 Cache and 2GB memory. The operating system is RedHat Linux version FC3, with kernel 2.6.9. The implementation of the MBLSOR is compiled with Intel Fortran compiler 8.0 version under the option -openmp. We use the sparse matrices generated from the following grid sizes: 2,250,000 grid blocks (equivalent to a grid of dimensions 150 x 150 x 100).

The time taken with the serial version MBLSOR is 1385 seconds while the parallel version takes 710 seconds. The speed-up ratio for this method in two processors is equal to 1.95. The actual speedup is smaller than the ideal speedup. This is likely caused by the fork-join overhead. We also measured the 2nd level cache load misses retired of serial LSOR and MBLSOR. The second example is from 300x100x300 grids. We can find that the 2nd level cache load misses retired are 61.9% and 35.1% respectively. From the above observation we find that data locality can help to improve the efficiency of LSOR.

5   Conclusions

In this paper, we described an efficient OpenMP implementation of a linear solver in black oil reservoir simulation system for SMPs. A novel MSLSOR method presented to improve data locality as well as parallelism, which is solution of system of banded linear equations. Simulation results show that the speed-up obtained is near the maximum speed-up, which indicates low synchronization overhead and good load balancing among processors.

Acknowledgements

The research is partially supported by the Key Technologies Research and Development Program of China under Grant No.2006038027015, the Hi-Tech Research and Development Program (863) of China under Grant No. 2006AA01Z105, Natural Science Foundation of China under Grant No.60373008, and by the Key Project of Chinese Ministry of Education under Grant No. 106019.

References

1. Zhlyuan, M., Fengjiang, J., Xiangming, X., Sunjiachang: Simulation of Black oil Reservoir on Distributed Memory Parallel Computers and Workstation Cluster. In: International meeting on petroleum engineering, Beijing, pp. 14–17 (1995)
2. Jiwu, S., Lizhong, G., Weisi, Z., Defu, Z.: Parallel computing of domain decomposition methods for solving numerical simulation of black of reservoir. Journal of Nanjing University 35(1) (January 1999)
3. OpenMP application program interface, ver 2.5, Tech. report (May 2005), http://www.openmp.org/
4. Xie, D., Adams, L.: New parallel method by domain partitioning. SIAM J. Sci. Comput. 27, 1513–1533 (2006)
5. Niethammer, W.: The SOR method on parallel computers. Numer. Math. 56, 247–254 (1989)
6. Evans, D.J.: Parallel SOR iterative methods. Parallel Computing 1, 3–18 (1984)
7. Hofhaus, J., Van de Velde, E.F.: Alternating-Direction Line Relaxation Methods on Multicomputers. SIAM Journal of Scientific Computing (2), Society of Industrial and Applied Mathematics, 454–478 (March 1996)

Parallel Data Flow Analysis for OpenMP Programs

Lei Huang, Girija Sethuraman, and Barbara Chapman

University of Houston, Houston TX 77004, USA
{leihuang,girija,chapman}@cs.uh.edu

Abstract. The paper presents a compiler framework for analyzing and optimizing OpenMP programs. The framework includes Parallel Control Flow Graph and Parallel Data Flow equations based on the OpenMP relaxed memory consistency model. It enables traditional compiler analyses as well as specific optimizations for OpenMP. Based on the framework, we describe dead code elimination and barrier elimination algorithms. An OpenMP code example is showed in the paper to illustrate the optimizations. The framework guarantees that the traditional optimizations can be performed safely to OpenMP programs, and it further increases the opportunities for more aggressive optimizations.

1   Introduction

OpenMP [7] and PThreads [2] are the most widely used programming models for parallelizing applications in shared memory systems. Whereas PThreads often requires major reorganization of a program's structure, the insertion of OpenMP directives is often straightforward. OpenMP directives impose a structured programming style with a simple means for synchronization that helps avoid some kinds of programming errors. A compiler translates an OpenMP code to threaded C/C++, Fortran code that will be linked with a thread library. It is easier for a compiler to analyze an OpenMP code than its corresponding threaded code due to its very structured style. However, most compilers do not exploit the fact to

```
                            mpsp_status = __ompc_single(__ompv_temp_gtid);
                            if(mpsp_status == 1)
#pragma omp single         {
{                             k = 1;
   k = 1;                  }
}                          __ompc_end_single(__ompv_temp_gtid);

if(k==1) ...               if(k==1) ...
```

(a)   An   OpenMP program with single construct

(b) The corresponding compiler translated threaded code

Fig. 1. A compiler translated OpenMP code

B. Chapman et al. (Eds.): IWOMP 2007, LNCS 4935, pp. 138–142, 2008.

analyze an OpenMP code. There is little or no optimization in most OpenMP compilers [8,6] before an OpenMP code is translated to a threaded code. Fig. 1 shows an OpenMP code and a compiler generated threaded code after the translation. Based on OpenMP semantics, it is known that k is equal to 1 after the OpenMP *single* construct. However, in the translated code, a compiler is not sure if k is equal to 1 or not at the *if(k==1)* statement since the value of *mpsp_status* is unknown at compile time.

2   OpenMP Memory Model

As described in [4] and OpenMP 2.5 specification, OpenMP is based on the relaxed consistency memory model, which means that each thread is allowed to have its own local view of shared data. The local value of a shared object may or may not be consistent unless there is a flush operation to force the value to be consistent over all threads. Besides the OpenMP flush directive, OpenMP synchronization mechanism (*omp barrier, omp critical, omp atomic, and locks*) contain implicit flush to make the data consistent. Most OpenMP constructs have implicit barriers at the end of them to ensure the synchronization of threads execution and keep the data consistent between them. An aggressive optimizing compiler can safely assume that there is no inter-thread data interactions until a flush operation has been reached. The OpenMP memory model simplifies the compiler analysis for parallel programs since a compiler can perform traditional analysis and optimizations safely between two synchronization operations. Therefore, most OpenMP compiler performs compiler optimizations after OpenMP has been translated to a threaded code, and limits the sequential analysis and optimizations to be performed only between two synchronizations.

3   Parallel Data Flow Analysis Framework for OpenMP

The idea of PCFG is similar to Program Execution Graph [1] and the Synchronized Control Flow Graph [3]. The distinction between our PCFG with them is that our PCFG is based on barrier and flush synchronizations, instead of event based synchronizations (such as post-wait). The PCFG is a directed graph (N, E, s, e), where N is the set of nodes including basic nodes, composite nodes, super nodes, and barrier nodes; E is a set of directed edges including sequential edges and parallel edges; s and e represent the entry and exit of a parallel region, respectively. A basic node is a basic block, or contains a **omp flush** directive. A composite node is composed of an OpenMP worksharing or synchronization construct and the basic nodes associated with it. A barrier node contains **omp barrier** directive only. A sequential edge indicates the sequential control flow within a thread. A parallel edge indicates a branch that more threads may take.

Fig 2 shows how the composite nodes containing worksharing directives are connected by parallel and sequential edges. In Fig 2 A and B, different threads may take different paths, so that parallel edges represent the branches for different threads. In Fig. 2 C, the *omp for* loop will be executed by all threads, and

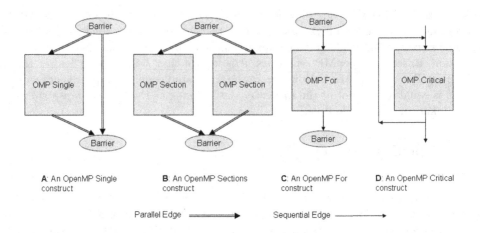

A: An OpenMP Single construct   **B**: An OpenMP Sections construct   **C**: An OpenMP For construct   **D**: An OpenMP Critical construct

Parallel Edge ⟹   Sequential Edge ⟶

Fig. 2. OpenMP Worksharing Constructs in PCFG

we use a sequential edge to connect it. Based on the *omp for* directive semantics, the enclosed loop should not have any data dependence. We treat the loop as a sequential loop in the PCFG. Fig. 2 D presents the PCFG for *omp critical* construct. A critical section is executed by multiple threads one by one, but never at same time. It is similar to the execution of a loop in a sequential program in terms of data propagation. We create a backward edge in the critical construct, so that a data defined in the critical region will be visible by next thread when it executes the critical section.

We introduce new equations in Parallel Data Flow Analysis(PDFA) to handle the distinctions between sequential and parallel data flow analysis as follows.

Super Node (S) Equations: A super node contains one or more composite nodes between two barrier nodes. The equation *In(s)* is the *Out(entry)* or *Out(Previous Barrier)*, and the Out(S) is a union of all composite nodes that directly reach the end of the super node. A flush set *Flush(S)* needs to be gathered in a super node, since a flush operation performs inter-thread data flow inside a super node.

$$Flush(S) = \bigcup_{b \in BasicNode(S)} Flush(b) \qquad (1)$$

We also need to compute the definitions that are killed in a super node. In a parallel program, if a definition has been killed in at least one of composite nodes inside a super node, it cannot reach the next super node. We need to perform a union operation for all must-be-killed definitions of all composite nodes in a super node. We define *MustKill(S)* of a super node to be a set of must-be-killed definitions in its all enclosed composite nodes.

$$MustKill(S) = \bigcup_{C \in CompositeNode(S)} MustKill(C) \qquad (2)$$

Composite Node (C) Equations: A composite node contains one or more basic nodes. We are interested in the intra-thread data flow, and the must-be-killed definitions in a composite node. The equation $In(C)$ is a union of $Out(C)$ of all previous composite nodes or $In(S)$ of its super node. $Out(C)$ is a union of its basic nodes that directly reach the end of the composite node. Intuitively, if a definition appears in the beginning of of a super node, but does not reach the end of a composite node inside the super node, it is included in the $MustKill(C)$ set.

$$MustKill(C) = In(S) - Out(C) \; (where \; S \; is \; the \; super \; node \; of \; C) \qquad (3)$$

Basic Node (b) Equations: We compute data flow equations for each basic node similarly with them in a sequential DFA. In addition, we need to handle inter-thread data flow by flush operations. We have the following modifications of equations for each basic node b:

$$Out(b) = (In(b) - Kill(b)) \bigcup Gen(n) \bigcup Flush(S)$$
$$(Where \; S \; is \; the \; super \; node \; enclosing \; b) \qquad (4)$$

$$Flush(b) = \begin{cases} \bigcup v & (v \in variables \; specified \; in \; the \; flush \\ & directive) \\ all \; shared \; variables & (if \; the \; flush \; does \; not \; specify \\ & any \; variables) \end{cases}$$
$$\qquad (5)$$

Barrier Node (Barrier) Equations: At a barrier point, threads wait until all threads reaching the point. It then flushes all shared variables before all threads execute the next super node. The $In(Barrier)$ equation is a union of all predecessors. And the $Out(barrier)$ set should exclude all definitions that must be killed in the previous super node.

$$Out(Barrier) = (In(barrier) - MustKill(S))$$
$$(where \; S \; is \; the \; previous \; super \; node \; of \; the \; barrier) \qquad (6)$$

3.1 Compiler Optimizations

Based on the above Parallel Control Flow Graph and Parallel Data Flow equations, we are able to perform traditional optimizations before an OpenMP code is lowered to a threaded code. A lowered threaded code may lose the structure of threads execution and interactions, and make it difficult for a compiler to optimize it globally. We can compute the data flow information such as reaching definitions based on the PDFA. Inter-thread and intra-thread use-definition and definition-use chain can be calculated and traditional optimizations such as copy propagation, dead code elimination, and partial redundancy elimination(PRE) etc. can be performed. Moreover, Barrier elimination is an optimization for parallel programs to remove redundant barriers so as to improve the performance. We can perform optimizations specific to parallel programs such as Barrier eliminataion based our PCFG and PDFA equations. Due to the page limit in this paper, we will present the algorithm of these optimizations in the future.

4   Conclusion and Future Work

The contribution of the paper is to present a compiler framework that enables high-level data flow analysis and optimizations for OpenMP by taking its semantics into consideration. The framework represents the intra- and inter-thread data flow in OpenMP based on the relaxed memory model. It enables classical global optimizations to be performed before an OpenMP code is lowered to a threaded code. Moreover, a compiler is able to perform more aggressive optimizations specific to OpenMP programs. In the future work, We will implement it into the OpenUH compiler [6] to further evaluate the work. We will explore more compiler optimizations based on the framework in cluster OpenMP implementation [5]. It could also be used in static analysis for detecting race conditions of an OpenMP program. We believe that the framework will lead to more aggressive optimizations and analysis for OpenMP.

References

1. Balasundaram, V., Kennedy, K.: Compile-time detection of race conditions in a parallel program. In: ICS 1989. Proceedings of the 3rd international conference on Supercomputing, pp. 175–185. ACM Press, New York (1989)
2. Buttlar, D., Nichols, B., Farrell, J.P.: Pthreads Programming. O'Reilly & Associates, Inc., Sebastopol (1996)
3. Callahan, D., Kennedy, K., Subhlok, J.: Analysis of event synchronization in a parallel programming tool. In: PPOPP 1990. Proceedings of the second ACM SIGPLAN symposium on Principles & practice of parallel programming, pp. 21–30. ACM Press, New York (1990)
4. Hoeflinger, J.P., de Supinski, B.R.: The openmp memory model. In: The 1st International Workshop on OpenMP (IWOMP 2005) (2005)
5. Huang, L., Chapman, B., Liu, Z.: Towards a more efficient implementation of OpenMP for clusters via translation to Global Arrays. Parallel Computing 31(10-12) (2005)
6. Liao, C., Hernandez, O., Chapman, B., Chen, W., Zheng, W.: OpenUH: An optimizing, portable OpenMP compiler. Concurrency and Computation: Practice and Experience, Special Issue on CPC 2006 selected papers (2006)
7. OpenMP: Simple, portable, scalable SMP programming (2006),
   http://www.openmp.org
8. Tian, X., Bik, A., Girkar, M., Grey, P., Saito, H., Su, E.: Intel OpenMP C++/Fortran compiler for hyper-threading technology: Implementation and performance. Intel Technology Journal 6, 36–46 (2002)

Design and Implementation of OpenMPD: An OpenMP-Like Programming Language for Distributed Memory Systems

Jinpil Lee, Mitsuhisa Sato, and Taisuke Boku

Department of Computer Science
Graduate School of Systems and Information Engineering
University of Tsukuba

Abstract. MPI is a de facto standard for parallel programming on distributed memory system although writing MPI programs is often a time-consuming and complicated work. We propose a simple programming model named OpenMPD for distributed memory system. It provides directives for data parallelization, which allow incremental parallelization for a sequential code as like as OpenMP. The result of evaluation shows that OpenMPD archieves 3 to 8 times of speed-up with a PC cluster with 8 processors with a small modification to the original sequential code.

1 Introduction

PC clusters are the typical platform for low cost high performance computing, and most of users code their program with MPI[6]. Since the MPI functions provide a variety of features to be applied to various parallelizing algorithm, it requires a complicated and tiresome parallel coding to users.

In this paper, we propose a new paradigm named OpenMPD based on Open-MPI [2] (caution: not "Open MPI", an MPI implementation), which is a conceptual model for easy MPI programming. OpenMPD which introduces a simple but effective features to describe typical scientific applications with a similar concept of OpenMP[5], a programming model on a shared memory system. OpenMPD provides OpenMP-like directives to describe array distribution and work sharing on the loop on parallel processes for data parallel programming.

Existing parallel programming languages or models for distributed memory system such as CAF (Co-Array Fortran)[3] or UPC (Unified Parallel C)[4] provide various features to describe parallel programs and to execute them in high efficiency. However, these features are complicated for most of users, and the implementation is invisible from them. It makes users difficult to write and optimize their code. On the designing of OpenMPD, we especially considered the simple and easily understandable concept of data parallel programming.

2 Overview of OpenMPD

The concept of Data parallel programming is relatively simple. Many applications can be parallelized by a few typical techniques. But MPI programming

B. Chapman et al. (Eds.): IWOMP 2007, LNCS 4935, pp. 143–147, 2008.
© Springer-Verlag Berlin Heidelberg 2008

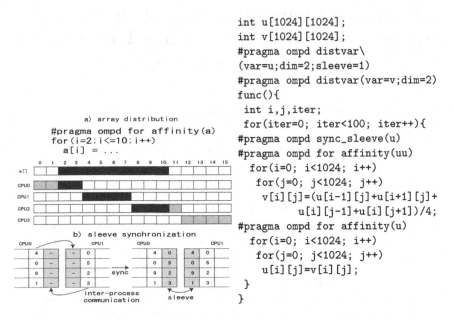

```
int u[1024][1024];
int v[1024][1024];
#pragma ompd distvar\
(var=u;dim=2;sleeve=1)
#pragma ompd distvar(var=v;dim=2)
func(){
  int i,j,iter;
  for(iter=0; iter<100; iter++){
#pragma ompd sync_sleeve(u)
#pragma ompd for affinity(uu)
  for(i=0; i<1024; i++)
   for(j=0; j<1024; j++)
    v[i][j]=(u[i-1][j]+u[i+1][j]+
        u[i][j-1]+u[i][j+1])/4;
#pragma ompd for affinity(u)
  for(i=0; i<1024; i++)
   for(j=0; j<1024; j++)
    u[i][j]=v[i][j];
  }
}
```

Fig. 1. Array distribution and sleeve syn-   **Fig. 2.** Program Example of OpenMPD
chronization

is often time-consuming because it requires a large modification of original code. OpenMPD is a language extension for programming on distributed memory systems to relieve users from such bothering MPI programming. Currently, C-language version of OpenMPD is only available. And the underlying parallel execution environment is MPI.

Programmer is responsible to discribe appropriate directives with their parameters to the original sequential code to parallelize it with implicit message passing features(currently, MPI). The Sequential code can be parallelized by a minimum modification. It makes parallel programming on distributed memory system more simple than message passing paradigm.

OpenMPD supports a typical parallelization based on data parallism, which is well used in varoius scientific applications. The following is brief explanation of data parallelization in OpenMPD, which can be described by directives.

- Because OpenMPD is implemented based on MPI, OpenMPD programs are excuted in SPMD(Single Program Multiple Data) model.
- OpenMPD supports global array distribution (supports block distribution, by directive **distvar**) and work sharing on for loop(directive **for**) which is basic features of data parallelism. **affinity** is an option to make work sharing assosiated to the array distribution. In consequence, each process on PC cluster refers a partial region of the array(see Fig.1.a).
- For synchronizaion of arrays, OpenMPD provides two kinds of methods. First one is, all-to-all data exchanging on parallel processes (array gathering, directive **gather**). This method is effective when you need random access to

the array. But it requres expensive communication cost at the same time. Second one is sleeve synchronization(directive **sync_sleeve**). "Sleeve" is a border region of a distributed array. In Fig.2, it is enough to exchange the one boundary elements(see Fig.1.b). Since this kind of synchronization appears frequently in scientific applications, OpenMPD provides the directive to describe it easily.

- Variables of basic types in C can be also synchronized by OpenMPD directives. Directive **sync_var** describes synchronization by broadcasting data from one process. Directive **reduction** describes collective communication among the processes. It can be used as an option of directive **for**.
- When the program can be devided into several functional block, some of them may be excuted independently. Directive **single** discribes *block-statement*(in C language) to be excuted on one process. By using **single** directive, functional independent *block-statements* can be excuted in parallel.

Fig.2 shows an typical program of OpenMPD solving a differential equation with a finite differencing method. The lines starting with **#pragma** are OpenMPD derectives to describe parallelization.

3   Implementation and Performance Evaluation

The compiler of OpenMPD is developed based on Omni OpenMP Compiler[7]. It translates the source code referring the directives to an MPI parallel code with supplement library function calls. At the code execution, these run-time library functions handle the distribution of array variables and inter-process communication with MPI functions. Therefore, the translated code and the run-time library functions are regarded as a complete MPI program with portability. And users can use MPI functions explicitly in OpenMPD code to optimize the performance.

Next, we show the performance of parallelized code by OpenMPD. We used a small Intel Xeon PC cluster with 8 nodes(1GB memory and 1000base-T Ethernet). The MPI library is LAM/MPI version 7.0.6.

We evaluated three benchmarks; NAS Parallel Benchmark[8] EP (NPB-EP) and CG (NPB-CG), and Himeno Benchmark[9] (HIMENO). These benchmark programs can be easily coded by OpenMPD in a simple domain decomposition manner.

Fig.3.a shows the speed-up of all benchmarks parallelized by OpenMPD and MPI(official version) when increasing the number of nodes. NPB-EP shows the excellent speed-up because there is no communication overhead, and OpenMPD handles it reasonably. For HIMENO, the speed-up from 4 to 8 nodes is degraded due to communication overhead and incomplete parallelization. Since the interconnection network of the tested PC cluster is Gigabit Ethernet, the communication performance is poor compared with the CPU power. We think the result is reasonable and we can achieve higher performance with more powerful interconnection network. The two benchmarks show almost the same spped-up compared with the MPI version. But NPB-CG shows worse performance as

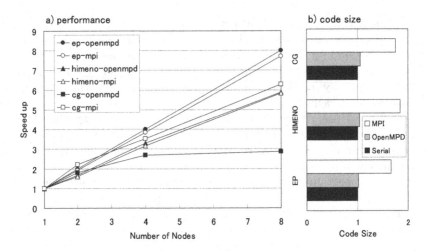

Fig. 3. Evaluation result

compared with other benchmark results. It is because that current implementation of OpenMPD does not support multi-dimensional distribution of array. We are now fixing the implementation to handle this problem. Fig.3.b shows the code size of OpenMPD and MPI version comparing with the serial code. It proves that OpenMPD needs very small modification to the serial code to parallelize.

Through the performance evaluation, we could confirm that the simple directives of OpenMPD are applicable for variety of codes without heavy effort and the compiler and functions of OpenMPD processing system work correctly.

4   Current Status and Future Work

Current implementation is very simple and we need to enhance it to apply our method to wider fields of applications and improve the performance and scalability. One of the interesting issue is a hybrid utilization of OpenMPD with OpenMP for easy and efficient hybrid programming to combine multi-thread and message passing features on SMP clusters.

References

1. Sato, M., Satoh, S., Kusano, K., Tanaka, Y.: Design of OpenMP Compiler for an SMP Cluster. In: Proc. of 1st European Workshop on OpenMP EWOMP 1999, pp. 32–39 (1999)
2. Boku, T., Sato, M., Matsubara, M., Takahashi, D.: OpenMPI - OpenMP like tool for easy programming in MPI. In: Proc. of 6th European Workshop on OpenMP (EWOMP 2004), pp. 83–88 (2004)
3. Co-Array Fortran, http://www.co-array.org/

4. Unified Parallel C, http://upc.gwu.edu/
5. OpenMP.org, http://www.openmp.org
6. The Message Passing Interface (MPI) standard,
   http://www-unix.mcs.anl.gov/mpi/
7. Omni OpenMP Compiler Project, http://phase.hpcc.jp/Omni/home.ja.html
8. NAS Parallel Benchmarks,
   http://www.nas.nasa.gov/Resources/Software-/npb.html
9. http://w3cic.riken.go.jp/E/HPC_e/HimenoBMT_e/index_e.html

A New Memory Allocation Model for Parallel Search Space Data Structures with OpenMP

Christophe Jaillet and Michaël Krajecki

CReSTIC SysCom, Université de Reims Champagne-Ardenne
Moulin de la Housse, BP 1039, 51687 Reims cedex 2
{christophe.jaillet,michael.krajecki}@univ-reims.fr

Abstract. OpenMP is a shared memory programming API (see [2] or
http://www.openmp.org), brought on UMA and CC-NUMA architectures, which supports multithreaded applications.

The present study deals with the management and arrangement of
memory, which may cause useless memory blocks reloading from the
shared memory to the processors caches. This occurs for some memory
access sensible applications, when programming with OpenMP: cache
faults may congest the system and induce a communication over-cost
between the processors. A new memory allocation model is presented,
enabling us to solve the memory contention phenomenon in the specific
case of shared memory programming with OpenMP.

Keywords: Parallel systems, memory allocation, shared memory,
OpenMP.

1 Introduction

Programming parallel applications in shared memory offers an effective solution
for the execution of applications that require significant computing efforts and
long computing times. These applications may suffer the consequences of an
execution times variability when programmed in OpenMP. This paper presents
this phenomenon and offers a low cost solution which, moreover, improves the
applications efficiency.

We currently are working in the field of combinatorial search and combinatorial optimization. The methods we implement [6] are based on the CSP formalism
and can be coarsely reduced to a tree search algorithm. The parallelization of the
applications results from a tasks generation, based on the tree development at a
chosen depth-level [3]. After the tasks generation, the performance of the parallel
applications is linked to three essential critical points: the tasks granularity, the
load balancing and the data locality.

The generated tasks obtained are relatively independent and can be treated
in any order, but are highly irregular (in computational time); thus a specific
dynamic load balancing strategy is necessary. The average granularity can be
chosen by fixing the tree development depth-level to any value, which essentially
depends on the size of the selected instance and the estimated execution time.

B. Chapman et al. (Eds.): IWOMP 2007, LNCS 4935, pp. 148–152, 2008.
© Springer-Verlag Berlin Heidelberg 2008

The applications developed are particularly linked to memory accesses, with quite few specific additional computation, so the memory accesses are essential for the performances: here is the core of the current study.

2   Computational Times Variability Using OpenMP

When executing our OpenMP applications in parallel, we can observe a high degree of variability of the computational times, especially with an increasing number of processors. *Tuning* the executions reveals a large number of processors synchronizations, caused by cache faults [4], that are not reproducible from one execution to the other.

This phenomenon is linked to the use of OpenMP since no such variability appears when programming in the same manner with pThreads.

It has to be noted that such a variability was observed on each of the architectures we worked on: SMP nodes (IBM SP3/SP4), SMP clusters (SunFire 6800), CC-NUMA machine (SGI Origin 3800, hypercube of SMP nodes).

Table 1. Computational times variability with OpenMP - L(2,14), Miller method, depth-level 5: best execution times; relative standard deviations (SGI Origin'3800). [*ClSe*: Client-server strategy – *SeI*: Server Initiated schedule].

procs	stq0	stq	mod0	mod	dyn0	ClSe	SeI
1	223,13	234,41	224,26	223,08	234,70	234,43	223,10
	0,1%	*0,0%*	*0,1%*	*0,1%*	*0,0%*	*0,0%*	*0,2%*
2	119,50	119,50	116,40	116,40	114,72	116,23	112,47
	24,9%	*34,5%*	*39,4%*	*42,2%*	*41,4%*	*40,0%*	*61,6%*
32	22,81	24,37	7,42	26,17	7,03	9,47	23,69
	47,1%	*30,4%*	*321,1%*	*27,5%*	*189,7%*	*40,2%*	*15,3%*
64	16,46	17,97	17,45	16,04	4,24	5,61	12,12
	27,6%	*21,6%*	*30,3%*	*26,0%*	*98,1%*	*48,1%*	*10,6%*

As an illustration, Table 1 gives computational times and relative standard deviations observed for 50 executions, solving the L(2,14) instance of the Langford problem [5] with a C/OpenMP program on a SGI Origin'3800 (R14000 processors, 500MHz, 500MB/proc.): it compares various native/hand-made parallel loop parallelization schemes. As the execution times were very irregular, we had to repeat the experiments from 20 to 50 or 100 times in order to measure the best observed time as a *normal conditions* execution time.

For 32 processors, for example in an OpenMP native modulo parallel loop, only 12 of the 50 executions gave a result close to the 7 seconds awaited time, whereas 13 of them lasted 22 to 26 seconds, and half of them consumed 55 to 65 seconds: the relative standard deviation observed is 320%.

2.1 Multithreaded Processes in Shared Memory

On a multiprocessor shared memory system, the threads execute themselves simultaneously on different processors, and communicate/synchronize through shared memory. Each thread can access the process's memory, with read/write asynchronous accesses, but may have a private storage place[1]. In the particular case of our applications, the threads use a significant amount of shared memory (for tasks descriptions storage), compute in their local storages, and use a small part of the shared memory to communicate (store/accumulate the tasks results).

On most multiprocessor systems, each processor disposes of its own cache memory unit. Assuming each thread is executed by a processor, it can be considered that each thread disposes of its own cache memory. The coherence of the global memory is ensured by caches coherence software/material systems [1], as illustrated by Fig. 1. The transfers to the caches are realized by *lines*.

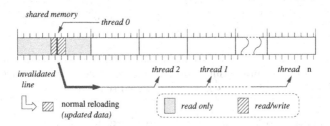

Fig. 1. Multithreaded process memory representation (virtual memory) - Cache memory management of shared memory

2.2 Execution Times Variability with OpenMP: A Solution

According to the POSIX standard, the memory spaces of any two threads are separated so they can not share data in a common cache line. As the high variability rate observed does not occur with pThreads, we proposed the hypothesis that the OpenMP memory pages manager enables the threads to share memory pages, as shown in Fig. 2.

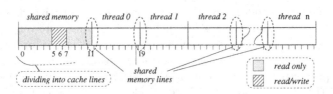

Fig. 2. OpenMP memory allocation scheme (conjecture): memory lines shared by the threads or with shared memory

---

[1] A part of the process's memory, entierly shared, kept as private for a given thread.

When a thread modifies the content of such a critical memory zone, it induces modifications of some cache line with data associated with another thread zone or with shared memory, and thus causes useless cache faults and cache lines reloading.

A new memory allocation model is proposed that avoids the different cache faults described above by using with free zones (unused memory segments) separating sensible memory blocks (see Fig. 3).

Using this new memory allocation model immediately corrected the execution irregularity, bringing cache faults rates are to the same level as with pThreads, hence limited to those necessary for the shared memory management. Added to the variability reduction, we even got a 3% improvement of the application performance.

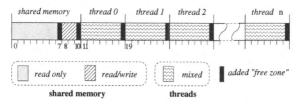

Fig. 3. Memory allocation model preventing useless cache faults, by separating the critical memory zones

3 Conclusion

In the field of combinatorial search/optimization, parallel applications use a large amount of shared memory and are very sensible to the memory management, especially for the cache memory accesses. We mentioned a high degree of computational times variability when programming with OpenMP, and proved that it is directly related to cache faults between the threads of the application process.

The solution proposed consists in adding some additional memory spaces in order to isolate the *sensible zones* from one another. It solves the problem without degrading the applications parallel performances.

We currently are working on a refained allocation model that may avoid any execution times variability for the parallel applications having an intensive use of the memory and of the data caches, and improve their performances.

Acknowledgement. Part of this work is supported by ROMEO II, regional high performance computating center of Champagne-Ardenne.

References

1. Adve, S., Adve, V., Hill, M., Vernon, M.: Comparison of Hardware and Software Cache Coherence Schemes. In: Proceedings of ISCA 1991, pp. 298–308. ACM, New York (1991)
2. Chandra, R., Dagum, L., Kohr, D., Maydan, D., McDonald, J., Menon, R.: Parallel Programming in OpenMP. Morgan Kaufmann, San Francisco (2000)

3. Habbas, Z., Krajecki, M., Singer, D.: Parallel Resolution of CSP with OpenMP. In: Proceedings of EWOMP 2000, pp. 1–8 (2000)
4. Itzkowitz, M., Wylie, B.J.N., Aoki, C., Kosche, N.: Memory Profiling using Hardware Counters. In: Proceedings of HPNC 2003, p. 17. ACM, New York (2003)
5. Jaillet, C., Krajecki, M.: Solving the Langford Problem in Parallel. In: Proceedings of ISPDC 2004, pp. 83–90. IEEE Computer Society, Los Alamitos (2004)
6. Krajecki, M., Jaillet, C., Bui, A.: Parallel Tree Search for Combinatorial Problems: a Comparative Study between OpenMP and MPI. Studia Informatica Universalis 4(2), 151–190 (2005)

Implementation of OpenMP Work-Sharing on the Cell Broadband Engine Architecture

Jun Sung Park, Jung-Gyu Park, and Hyo-Jung Song

Samsung Electronics

Abstract. The Cell Broadband Engine (CBE) is a single-chip multi-processor composed of one Power Processor Element (PPE) core and multiple Synergistic Processor Element (SPE) cores. Due to the heterogeneous processor type of the CBE, multi-core programming for the CBE is very difficult which can be alleviated by OpenMP. In this paper, we introduce some issues in implementing OpenMP work-sharing constructs on the CBE. To address overheads in creating and managing work-sharing control blocks across different processor types, we have come up with a method for aggressively reusing the work-sharing control block by precisely checking the reusability of it. This scheme allows us better performance over naive implementation of work-sharing constructs, when evaluated with modified EPCC benchmark.

1 Introduction

The Cell Broadband Engine (CBE) [1] is relatively new architecture which targets intensive computation such as multimedia streaming and game applications, as it is used in PlayStation 3. It can provide both performance and flexibility for wide range of devices including HDTV sets and home servers by allowing downloadable software module (i.e. codec) running as fast as hardware module. However, implementing software on a CBE processor requires a good understanding of the processor architecture. OpenMP can be one candidate to reduce these efforts. Because of hardware features of the CBE, implementing OpenMP for the CBE should be carefully considered to get utmost performance.

1.1 The Cell Broadband Engine Architecture

The CBE architecture is a heterogeneous multiprocessor composed of one 64-bit Power Processor Element (PPE) and eight specialized coprocessors called Synergistic Processor Elements (SPEs), interconnected by a high-bandwidth bus interface. The PPE runs operating system, manages main memory, and coordinates the SPEs. The SPE has a processor unit capable of SIMD computation, 256-KByte local-store memory (LS) for program and data, and a memory flow controller (MFC) which allows DMA to the main memory. Communication overheads between processing elements (i.e. synchronization and main memory access), although they are connected over high-bandwidth bus, are in no way negligible.

B. Chapman et al. (Eds.): IWOMP 2007, LNCS 4935, pp. 153–156, 2008.

Barrier Synchronization. The communication mechanisms such as mailboxes and DMA transfers are available to implement the synchronization among SPE threads [2]. Supporting more-complex synchronization mechanisms is a set of DMA atomic functions [3]. The latency of the DMA transfer is approximately 100-200 SPU cycles [3] and this high latency becomes an issue of barrier synchronization among SPE threads.

PPE Serviced SPE Library Function. Since the SPE is designed only for computation-intensive workloads, it should delegate I/O or main memory allocation to the PPE. This feature is provided as PPE-serviced function [4] that allows SPE program to seamlessly use PPE-side services. For example, if program code in a SPE invokes file I/O API, it is redirected to PPE and the result is returned to the SPE code. Although it is easy to use, it is not free of overheads such as DMA and activation (and context switch) of PPE threads to get the service done. According to our benchmark evaluation, the latency of PPE-serviced malloc function takes more than 100 times of that of DMA operation.

1.2 Problem Description

For each work-sharing construct, a control block is necessary to maintain related data structure of runtime library and to be shared among team threads. With NOWAIT clause, the implied barrier of work-sharing construct is disabled and multiple work-sharing constructs can be alive simultaneously [5]. The simplest solution for handling control blocks for these is to create each instance for each work-sharing construct when necessary. But frequent allocation and deallocation of memory can be detrimental to the runtime performance. Furthermore, it is impossible to pre-allocate the instances because the exact number of work-sharing constructs that may be alive simultaneously in a parallel region is only known at runtime. To reduce the overhead of this dynamic memory allocation, [6] proposed a method to allocate a chunk of work-sharing control blocks at once and keep in a queue. This scheme can reduce the overhead in frequent memory allocation but the reuse of the control blocks is available only after the barrier is encountered. When the queue becomes full, barrier synchronization should be inserted intentionally to prevent the overhead in resizing the pool. In the CBE with team threads running on SPEs, every work-sharing control block is created and stored in the main memory. For this to happen, a SPE thread uses PPE-serviced malloc which has quite heavy performance overheads as described in previous section. Although this dynamic memory allocation problem can be solved by pre-allocating a very large size array of work-sharing control block on the main memory, this is space-inefficient and overheads of memory allocation and initialization that are to be paid on every parallel region.

2   Proposed Work-Sharing Implementation in the CBE

The key idea of proposed scheme is to maximize the reuse of work-sharing control blocks by checking its reusability. A container, which is the unit of reuse

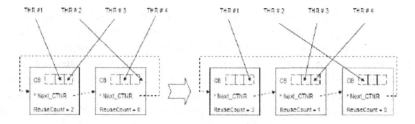

Fig. 1. Work-sharing containers in circular list

in our scheme, consists of a chunk-sized array of work-sharing control blocks, a reusability counter, and a pointer to the next container instance. In our implementation for the CBE, master thread on PPE creates the first container in the beginning of parallel region and team threads on each SPE execute work-sharing constructs referring the work-sharing control blocks in this container. After a team thread finishes referring all the work-sharing control blocks in a container, it increases the reusability counter by one and try to refer the next work-sharing control block in the next container. If it is the first thread to refer the next container, then checks the reusability counter of the next container. If the value is equal to the number of team threads, it resets counter value to zero and reuses the next container. Otherwise, it creates new container to insert in a circular list of containers and sets this as the next container. Therefore, by using this scheme, no barrier synchronization is needed to reuse the work-sharing control blocks and the overhead of memory allocation is reduced efficiently.

3   Experiments

We chose the dynamic scheduling testcase from the EPCC benchmarks suite [7] and the performance is evaluated on Sony Playstation3. We ran the testcase with 4 SPE threads as team threads and set the array size of work-sharing control

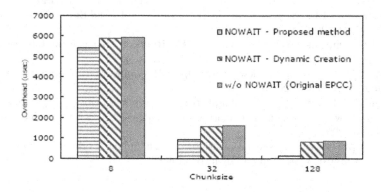

Fig. 2. Overhead (usec) of dynamic scheduling test of EPCC with 4 SPE threads

blocks in a container to four. To understand the overhead of barrier synchronization as a performance factor in CBE, the first step of our analysis is measuring the original benchmark test which has an implicit barrier at the end of the work-sharing construct. Secondly, we measured two work-sharing construct management schemes: proposed circular list scheme and dynamic work-sharing control block creation scheme. Since there is no testcase for multiple alive work-sharing construct case in EPCC benchmarks, we prepared *modified* dynamic scheduling testcase by adding NOWAIT clause in the code to eliminate the implied barrier synchronization. Figure 2 shows the measured overhead against chunk size for the dynamic scheduling test and *modified* test. From the observation of results, our proposed scheme provides better performance than dynamic creation scheme at each chunk size since the overhead of PPE-serviced malloc function is reduced dramatically. And also, we can see the overheads from the dynamic creation scheme are nearly equal to those of the implied barrier synchronization for all work-sharing constructs. NOWAIT clause to avoid the barrier synchronization overhead is not effective for the dynamic creation scheme.

4   Conclusion

In this paper, our approach to reduce the runtime overhead gives better performance by aggressively reusing the work-sharing control blocks. This is effective especially when the application executes work-sharing constructs without implicit or explicit barrier synchronization on the CBE.

References

1. Pham, D., et al.: The design and implementation of a first-generation CELL processor. In: IEEE International Solid-State Circuites Conference (2005)
2. Zhao, Y., Kennedy, K.: Dependence-based Code Generation for a CELL Processor. In: International Workshop on Languages and Compilers for Parallel Computing (2006)
3. Kiestler, M., Perrone, M., Petrini, F.: The design and implementation of a first-generation CELL processor. In: IEEE International Solid-State Circuites Conference, pp. 26–28 (2005)
4. IBM Systems and Technology Group: SPE Runtime Management Library Version 2.0, pp. 59–62 (2006)
5. OpenMP Architecture Review Board: OpenMP Application Program Interface Version 2.5, pp. 32–46 (2005)
6. Zhang, G., Silvera, R., Archambault, R.: Structure and algorithm for implementing OpenMP workshares. In: The Workshop on OpenMP Applications and Tools, pp. 110–120 (2004)
7. Bull, J.M.: Measuring synchronization and scheduling overheads in OpenMP. In: First European Workshop on OpenMP (1999)

Toward an Automatic Code Layout Methodology

Joseph B. Manzano[1], Ziang Hu[1], Yi Jiang[1], Ge Gan[1],
Hyo-Jung Song[2], and Jung-Gyu Park[2]

[1] Department of Electrical and Computer Engineering, University of Delaware,
Newark, Delaware 19716, USA
[2] Software Laboratories, Corporate Technology Operations, Samsung Electronics
{jmanzano,hu,jyi,gan}@capsl.udel.edu,
{hjsong,junggyu.park}@samsung.com

Abstract. This paper presents a study on an automatic code layout methodology for multi core architectures with explicit memory hierarchies. Code layout techniques are employed to run large programs on such systems. This study shows the effects of different buffer schemes and replacement policies on a set of benchmarks. These schemes are considered to be more flexible than current approaches. Moreover, these schemes can be used as a foundation to build frameworks for high level parallel programming models in such multi core architectures. The current work has been implemented in IBM's Cell Broadband Engine, but it can be extended to similar architectures.

1 Introduction

This paper presents a software-based code overlay framework that can be used to implement high level parallel programming models. Such approach an is more flexible than current approaches - since parameters, like replacement policies and buffer behaviors, can be easily changed. The current study is implemented on IBM's CBE architecture [1]. Even though this problem is not new (many studies have been conducted in this area before the advent of virtual memory and in early Operating Systems[1]), the resurface of heterogeneous architectures has given a new life to this problem.

The main objective of this paper is to study the following problem: given a fixed size of local memory, decide a code buffering behavior and replacement policy, such that the overall execution time for a given application is minimized. The paper is divided as follows. Section 2 talks about the general methodology and framework of the software based code layout approach. Section 3 provides the experimental test bed and the results obtained from this test bed. Finally, section 4 provides conclusions and future work.

2 Framework and Methodology

A partition is a set of function calls that will be loaded as a single entity during execution. Partitions can be created by the user (using special pragma directives)

---

[1] Like the infamous OVL files in Microsoft Disk Operating System.

B. Chapman et al. (Eds.): IWOMP 2007, LNCS 4935, pp. 157–160, 2008.

or automatically by the compiler. A pointer to each partition and its size are saved in a list, called the partition list. The index in this list is used to load and to identify partitions for several purposes. This index is assigned to the partition during link time and embedded in any call which target is inside a partition. The static libraries (such as libc) and the program entry point always reside in local memory. For this reason, they are referred as un-partitioned code.

Under the code layout framework, a function can be called in two ways. If the function resides in local memory then the function call proceeds as usual. If the function is in a partition then the partition manager is called with certain extra information about the partition being called. The partition manager is a small ghost function, whose purpose is to load a code partition when needed and to create or to restore the state that is needed by such a partition. To maintain the state of old partitions in function call chains, a special structure needs to be maintained. This structure is called the partition stack and it is analogous to a normal function stack. It keeps the caller partition id, return address and other useful information for each partition that has been called. The stack will be operated on at the beginning and at the end of every partition manager call.

When a partition is loaded to local memory, the place in which it resides is called the partition buffer. This buffer can be assigned many different partitions during runtime. Furthermore, it can be subdivided into segments by the programming environment. The buffer exhibits several behaviors and replacement policies. The behavior presented in this paper is cache like and it exhibits two replacement policies. One of them is called the modulus replacement policy. In this approach, a partition is replaced in a modulus fashion. Its main disadvantage is that some important partitions might be evicted unnecessarily. The other policy is called Least Recently Used (LRU) and it shares similar aspects with its cache related cousin. In this case, each buffer will be assigned a number (time to live) which is decremented every time that a partition is replaced in another buffer. The buffers with the minimum time to live are targeted to be replaced in the next load.

In this study, the buffer had three implementations that exhibit the former behaviors and replacement policies. The first one is called the single buffer scheme. It was designed as a proof of concept for code overlay. Its buffer behavior and replacement policy are trivial (i.e. always replace). The second implementation is the double buffer scheme. In this scheme, the buffer is divided into two. The replacement scheme in this approach is modulus. Finally, in the N-buffer scheme, the buffer can be subdivided into n equal sized sub-buffers. The replacement policies in this scheme are modulus and LRU. Results for each implementation can be seen in the next section.

3   Experimental Test Bed

In this study, a Playstation[2] 3 with Yellow Dog Linux installed was used. The group of programs that were used as a test bed is given in table 1. The metrics

---

[2] Playstation is a trademark of Sony Inc.

Table 1. Applications Test Bed

Application	Description
DSP	A set of DSP kernels.
GZIP	De/Compression utility.
Jacobi	An application of Jacobi method.
Laplace	An application of the Laplace transform.
MD	A simple molecular dynamic application.
MGRID	A Multi-Grid solver.

used in these experiments are the number of DMA transfers and the execution time for each program. Figure 1 depicts the number of DMA required for selected applications when using 1 through 4 sub-buffers. Figure 2 shows the runtime of the selected applications when running on single and double buffer schemes. Finally, figures 3 and 4 show the number of DMA transfers and execution times of the LRU and the modulus policies with four buffers.

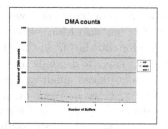
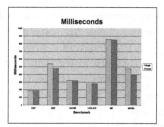

Fig. 1. DMA Transfers: 1, 2, 3 & 4 buffs   **Fig. 2.** Execution time: single & double

Figure 1 shows that extra buffers do not necessarily represent an increase in performance (or a reduction in DMA transfers). More buffers are useful when more partitions are available and when they are called often. The biggest advantage seen in figure 1 is the synthetic case 7. The sharp decrease in DMA transfers is expected since the test case was designed to test function calls between partitions in loops. The GZIP application shows a dramatic decrease between one and two buffers, but only minor decreases (18 DMA transfer from 2 to 3 buffers and 3 DMA transfers from 3 to 4 on average) when the number of buffers are increased. The MGRID application shows a linear decrease that implies that more buffers can improve its performance even further. These results support the idea that the number of sub-buffers is highly dependent on the application's structure.

Figures 2 and 4 show the execution time of six applications from which two show great variations. These graphs show a strong correlation between DMA and execution time and how extra DMA transfers can affect the performance of a given application, anything from 11 to 30 percent compared with a lower

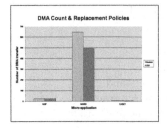

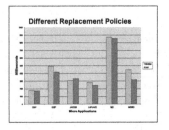

Fig. 3. LRU & MOD policies: DMA   **Fig. 4.** LRU & MOD policies: Time

performance schemes. Figures 3 and 4 show that the LRU policy can be advantageous in applications with high number of partitions, like MGRID. However, it does show some performance degradation for applications when the number of partitions is low. The greatest of these performance degradations is 8 % for the JACOBI application, which has only one partition in the whole program.

4   Conclusions and Future Work

This paper presents an empirical analysis for a group of buffer behaviors and replacement policies for software code layout. IBM has also been aware of the code layout problem and it provided some method for code layout in their CELL architecture [3]. However, this method is too strict since it depends on GCC's overlays [2]. However, the test-bed presented here is indeed small and part of the future work is to extend it to include more test cases. Work on a Prefetching scheme is also planned. This scheme depends on a special structure called the Partition Graph, which illustrates the relationships between partitions. Giving this graph, many parameters, like partition dependencies, usage and an optimal number of sub-buffers, can be calculated and used to pre-fetch partitions. Finally, improvements, such as dynamically switching between modulus and LRU policies, are planned.

Acknowledgments

I would like to thank ET International for their support of this project. I would also like to thank my adviser, Guang R. Gao, for all his advice and guidance and my co worker Ioannis E. Venetis for his help in reviewing this paper.

References

1. IBM Research Overview of the Cell Broadband Engine Processor. Cell Broadband Engine Programming Handbook. Version 1.0, pp. 34–44 (2006)
2. Chamberlain, S., Taylor, I.L.: Linker Scripts: Overlays. Using LD, The GNU linker. pp. 41–42
3. IBM Research Compiler and Runtime Support for Code Partitioning. Cell Broadband Engine Programming Handbook. Version 1.0, pp. 616–617 (2006)

An Efficient OpenMP Runtime System
for Hierarchical Architectures

Samuel Thibault, François Broquedis, Brice Goglin,
Raymond Namyst, and Pierre-André Wacrenier

INRIA Futurs - LaBRI
351 cours de la libération
33405 Talence cedex, France
{thibault,goglin,namyst,wacrenier}@labri.fr,
francois.broquedis@etu.u-bordeaux1.fr

Abstract. Exploiting the full computational power of always deeper hierarchical multiprocessor machines requires a very careful distribution of threads and data among the underlying non-uniform architecture. The emergence of multi-core chips and NUMA machines makes it important to minimize the number of remote memory accesses, to favor cache affinities, and to guarantee fast completion of synchronization steps. By using the BubbleSched platform as a threading backend for the GOMP OpenMP compiler, we are able to easily transpose affinities of thread teams into scheduling hints using abstractions called bubbles. We then propose a scheduling strategy suited to nested OpenMP parallelism. The resulting preliminary performance evaluations show an important improvement of the speedup on a typical NAS OpenMP benchmark application.

Keywords: OpenMP, Nested Parallelism, Hierarchical Thread Scheduling, Bubbles, Multi-Core, NUMA, SMP.

1 Introduction

The emergence of deeply hierarchical architectures based on multi-threaded multi-core chips and NUMA machines raises the need for a careful distribution of threads and data. Indeed, cache misses and NUMA penalties become more and more important with the complexity of the machine, making these constraints as important as parallelization. They require some new programming models and new tools to make the most out of these underlying architectures.

As quoted by Gao *et al.* [GSS+06], it is important to expose domain-specific knowledge semantics to the various software components in order to organize computation according to the application and architecture. Indeed, the whole software stack, from the application to the scheduler, should be involved in the parallelizing, scheduling and locality adaptation decisions by providing useful information to the other components.

Therefore, in OpenMP frameworks, the information extracted by the compiler (about memory affinity and adherence to the same parallel section) can be

B. Chapman et al. (Eds.): IWOMP 2007, LNCS 4935, pp. 161–172, 2008.
© Springer-Verlag Berlin Heidelberg 2008

very useful for the guidance of task/thread scheduling. On the other hand, it is very important to rely on architecture specific constraints when making these scheduling decisions. A tight interaction between the OpenMP stack and the underlying hardware-aware scheduler is thus required.

The most delicate point, when dealing with irregular applications, is to exploit this knowledge at runtime (during the whole execution time) so as to maintain a good balancing of threads when events arise (task termination, creation of new embedded parallel sections, blocking synchronization, etc.).

In this paper, we propose a hierarchical threading library able to follow/obey scheduling directives and advices in a very powerful manner. Scheduling information (affinity, group membership) is attached to bubbles, which are abstractions that can recursively group threads or bubbles sharing common properties.

We report on preliminary experiences on top of a 8-way multi-core NUMA machine and we show that running OpenMP applications on top of our runtime system greatly enhances performance on hierarchical architectures under irregular conditions. We also propose insights regarding the extraction of useful information by the compiler for our runtime and discuss the addition of a couple of non-standard OpenMP directives that would improve performance.

2   Scheduling Applications Featuring Nested, Irregular Parallelism

Achieving the best possible performance when programming OpenMP applications requires developers to expose the parallelism and to explicitly design their code to drive its parallel behavior. Therefore, it is quite common nowadays to define per-thread specific data structures (in order to avoid false-sharing) and use a static, possibly pre-calculated, distribution of the workload to get good data locality [MM06]. Indeed, this model suits very well regular applications with coarse-grain parallelism.

However, this approach is hardly usable when dealing with irregular applications that rather need a dynamic load balancing mechanism. The use of complex synchronization schemes, or even blocking systems calls, may also be responsible for introducing irregularities regarding the computing load on the available processors. Using OpenMP dynamic scheduling directives can sometimes improve performance. In some cases, however, it may penalize data locality or even introduce false sharing effects, which can severely impact performance on hierarchical architectures.

Another approach is to increase the number of potential parallel tasks using nested parallelism, so that threads can be dynamically (re)allocated according to the workload disparity. The performance of such a dynamic thread management, when supported[1], heavily relies on the underlying runtime implementation, but also on the underlying operating system's scheduler. This explains why OpenMP users have been experiencing poor performance with the nested capabilities of

---

[1] Nested parallelism is currently an optional feature in OpenMP.

some OpenMP compilers, and have ended up performing explicit thread programming on top of OpenMP [BS05, GOM+00] or explicitly binding thread groups to processors [Zha06].

Nevertheless, there exists some very good implementations of OpenMP nested parallelism, such as Omni/ST [TTSY00] for instance. Such implementations are typically based on a fine-grain thread management system that uses a fixed number of threads to execute an arbitrary number of *filaments*, as done in the Cilk multithreaded system [FLR98]. The performance obtained over symmetrical multiprocessors is often very good, mostly because many tasks can be executed sequentially with almost no overhead when all processors are busy. However, since these systems provide no support for attaching high level information such as memory affinity to the generated tasks, many applications will actually achieve poor performance on hierarchical, NUMA multiprocessors.

One could probably enhance these OpenMP implementations to use affinity information extracted by the compiler so as to better distribute tasks or threads over the underlying processors. However, since only the underlying thread scheduler has complete control over scheduling events such as processor idleness, blocking syscall or even thread preemption, this information could only be used to influence task allocation at the beginning of each parallel section.

We believe that a better solution would be to transmit information extracted by the compiler to the underlying thread scheduler *in a persistent manner*, and that only a tight integration of application-provided meta-data and architecture description can let the underlying scheduler take appropriate decisions during the whole application run time. In other words, one can see this configurable scheduler framework as a domain-specific language enabling scientists to transfer their knowledge to the runtime system [GSS+06].

3   MaGOMP: An Implementation of GNU OpenMP for Hierarchical Machines

To evaluate the potential gain of providing a thread scheduler with persistent information extracted by an OpenMP compiler, we have extended the GNU OpenMP runtime system (i.e. the `libgomp` library) so as to rely on the *Marcel* thread library. This library provides facilities for attaching various information to groups of threads, together with a framework that helps to develop schedulers capable of using these metadata. Scheduling policies are simply developed as *plug-ins*.

Before describing our extensions to the GNU OpenMP compiler suite, we first present the most important features of the *Marcel* library.

3.1   The *Bubble* Scheduling Model

Marcel is a POSIX-compliant thread library featuring extensions for easily writing efficient, customized schedulers for hierarchical architectures. The API of *Marcel* provides functions to group threads using nested sets called **bubbles** [Thi05]. These abstractions allow programmers to model the relationships between the different threads of an application. Figure 1 illustrates this concept:

Fig. 1. Expressing thread relationships: graphical and tree-based representations

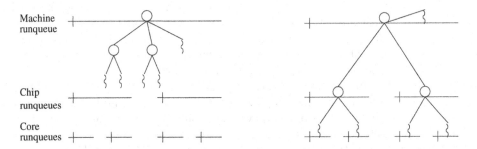

Fig. 2. Scheduling of bubbles and threads on the runqueues of a hierarchical machine

four threads are grouped as pairs in bubbles (assuming they work on the same data), which are themselves grouped along another thread in a larger bubble (assuming they share information less often). Bubbles allow expression of relationships like data sharing, collective operations, or more generally a particular scheduling policy need (serialization, gang scheduling, etc.). Hierarchical machines are modelled with a hierarchy of runqueues. Each component of each hierarchical level of the machine is represented by one runqueue: one per logical processor, one per core, one per chip, one per NUMA node, and one for the whole machine. *Marcel*'s ground scheduler then uses a hierarchical *Self-Scheduling* algorithm. Whenever idle, a processor scans all runqueues that span it, and executes the first thread that is found, from bottom to top. For instance, if the thread is on a runqueue that represents a chip, it may be run by any processor of this chip (see Figure 2).

As mentioned previously, *Marcel* provides a high-level API for writing powerful and portable schedulers that manipulate threads, bubbles and runqueues. Threads and bubbles are equally considered as **entities**, while bubbles and runqueues are equally considered as **scheduling holders**, so that we end up with entities (threads or bubbles) that we can schedule on holders (bubbles or runqueues). Primitives are then provided for manipulating entities in holders. Runqueues can be accessed through vectors, and can be walked through thanks to "parent" and "child" pointers. Some functions permit to gather statistics about bubbles so as to take appropriate decisions. This includes for instance the total number of threads and the number of running threads, but also various information such as the accumulated expected and current CPU computation time or memory usage, or the cache miss rates.

Writing a high-level scheduler actually reduces to writing some hook functions. The main one is actually called when the ground Self-Scheduler encounters a bubble during its search for the next thread to execute. The default implementation just looks for a thread in the bubble (or one of its sub-bubbles) and switches to it. The `bubble_tick()` hook is called when some time-slice for a bubble expires, and hence permits periodic operations on bubbles with a per-bubble notion of time. Of course, mere "daemon" threads can also be started for performing background operations. As a result, scheduling experts may manipulate threads with a high level of abstraction by deciding the placement of bubbles on runqueues, or even temporarily putting some bubbles aside (by defining their own runqueues that the basic Self-Scheduler will not look at).

3.2   Generating Bubbles Out of OpenMP Parallel Sections

The GNU OpenMP compiler[gom], GOMP, is based on an extension of the GCC 4.2 compiler that converts OpenMP pragmas into threading calls. The creation of threads and teams is actually delegated to a shared library, `libgomp`, which contains an abstraction layer to map OpenMP threads onto various thread implementations. This way, any application previously compiled by GOMP may be relinked against an implementation of `libgomp` on another thread type and transparently work the same.

We used this flexible design to develop MaGOMP, a port of GOMP on top of the *Marcel* threading library in which BubbleSched is implemented. To do so, a *Marcel* adaptation of `libgomp` threads has been added to the existing abstraction layer. We rely on *Marcel*'s fully POSIX compatible interface to guarantee that MaGOMP will behave as well as GOMP on pthreads. Then, it becomes possible to run any existing OpenMP application on top of BubbleSched by simply relinking it.

Once *Marcel* threads are created they basically behave by default as native pthreads without any notion of team or memory affinity. BubbleSched hooks have been added in the `libgomp` code to provide information about thread teams by creating bubbles accordingly.

Therefore, when a thread encounters a nested parallel region and becomes the master of a new team, it creates a bubble within its currently holding bubble. Then, it moves itself into this new bubble and creates the team's slave threads inside it. Finally, the master dispatches the workload across the team. Once their work is completed, slave threads die while the master destroys the bubble and returns to its original team. As shown on Figure 3, only a few lines of code are needed to associate a nested team hierarchy with a bubble hierarchy.

3.3   A Scheduling Strategy Suited to OpenMP Nested Parallelism

The challenge of a scheduler for the nested parallelism of OpenMP resides in how to distribute the threads over the machine. This must be done in a way that favors both a good balancing of the computation and, in the case of multicore and NUMA machines, a good affinity of threads, for better cache effects and avoiding the remote memory access penalty.

```
void gomp_team_start (void (*fn) (void *), void *data, unsigned nthreads,
                      struct gomp_work_share *work_share) {
  struct gomp_team *team;
  team = new_team (nthreads, work_share);
  ... /* Pack 'fn' and 'data' into the 'start_data' structure */

  if (nthreads > 1 && team->prev_ts.team != NULL) {
    /* nested parallelism, insert a marcel bubble */
    marcel_bubble_t *holder = marcel_bubble_holding_task (thr->tid);
    marcel_bubble_init (&team->bubble);
    marcel_bubble_insertbubble (holder, &team->bubble);
    marcel_bubble_inserttask (&team->bubble, thr->tid);
    marcel_attr_setinitbubble (&gomp_thread_attr, &team->bubble);
  }

  for(int i=1; i < nbthreads; i++) {
      pthread_create (NULL, &gomp_thread_attr,
                      gomp_thread_start, start_data);
      ...
  }
}
```

Fig. 3. One-to-One correspondence between *Marcel*'s bubble and GOMP's team hierarchies

For achieving this, we wrote a **bubble spread** scheduler consisting of a mere recursive function that uses the API described in section 3.1 to greedily distribute the hierarchy of bubbles and threads over the hierarchy of runqueues. This function takes in an array of "current entities" and an array of "current runqueues". It first sorts the list of current entities according to their computation load (either explicitly specified by the programmer, or inferred from the number of threads). It then greedily distributes them onto the current runqueues by keeping assigning the biggest entity to the least loaded runqueue[2], and recurse separately into the sub-runqueues of each current runqueue.

It often happens that an entity is much more loaded than others (because it is a very deep hierarchical bubble for instance). In such a case, a recursive call is made with this bubble "exploded": the bubble is removed from the "current entities" and replaced by its content (bubbles and threads). How big a bubble needs to be for being exploded is a parameter that has to be tuned. This may depend on the application itself, since it permits to choose between respecting affinities (by pulling intact bubbles as low as possible) and balancing the computation load (by exploding bubbles for having small entities for better distribution).

This way, affinities between threads are taken into account: since they are by construction in the same bubble hierarchy, the threads of the same external loop

---

[2] This algorithm comes from the greedy algorithm typically used for resolving the bi-partition problem.

iterations are spread together on the same NUMA node or the same multicore chip for instance, thus reducing the NUMA penalty and enhancing cache effects.

Other repartition algorithms are of course possible, we are currently working on a even more affinity-based algorithm that avoids bubble explosions as much as possible.

4   Performance Evaluation

We validated our approach by experimenting with the BT-MZ application. It is one of the 3D Fluid-Dynamics simulation applications of the Multi-Zone version of the NAS Parallel Benchmark [dWJ03] 3.2. In this version, the mesh is split in the x and y directions into zones. Parallelization is then performed twice: simulation can be performed rather independently on the different zones with periodic face data exchange (coarse grain *outer* parallelization), and simulation itself can be parallelized among the z axis (fine grain *inner* parallelization). As opposed to other Multi-Zone NAS Parallel Benchmarks, the BT-MZ case is interesting because zones have very irregular sizes: the size of the biggest zone can be as big as 25 times the size of the smallest one. In the original SMP source code, outer parallelization is achieved by using Unix processes while the inner parallelization is achieved through an OpenMP static parallel section. Similarly to Ayguade *et al.* [AGMJ04], we modified this to use two nested OpenMP static parallel sections instead, using $n_o * n_i$ threads.

The target machine holds 8 dual-core AMD Opteron 1.8GHz NUMA chips (hence a total of 16 cores) and 64GB of memory. The measured NUMA factor between chips[3] varies from 1.06 (for neighbor chips) to 1.4 (for most distant chips). We used the class A problem, composed of 16 zones. We tested both the Native POSIX Thread Library of Linux 2.6 (NPTL) and the *Marcel* library, before trying the *Marcel* library with our *bubble spread* scheduler.

We first tried non-nested approaches by only enabling either outer parallelism or inner parallelism, as shown in Figure 4:

Outer parallelism($n_o * 1$): Zones themselves are distributed among the processors. Due to the irregular sizes of zones and the fact that there is only a few of them, the computation is not well balanced, and hence the achieved speedup is limited by the biggest zones.

Inner parallelism($1 * n_i$): Simulation in zones are performed sequentially, but simulations themselves are parallelized among the z axis. The computation balance is excellent, but the nature of the simulation introduces a lot of inter-processor data exchange. Particularly because of the NUMA nature of the machine, the speedup is hence limited to 7.

So as to get the benefits of both approaches (locality and balance), we then tried the nested approach by enabling both parallelisms. As discussed by

---

[3] The NUMA factor is the ratio between remote memory access and local memory access times.

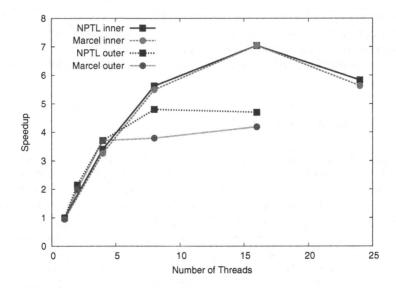

Fig. 4. Outer parallelism ($n_o * 1$) and inner parallelism ($1 * n_i$)

DURAN *et al.* [DGC05], the achieved speedup depends on the relative number of threads created by the inner and the outer parallelisms, so we tried up to 16 threads for the outer parallelism (i.e. the maximum since there are 16 zones), and up to 8 threads for the inner parallelism. The results are shown on Figure 5. The nested speedup achieved by NPTL is very limited (up to 6.28), and is actually worse than what pure inner parallelism can achieve (almost 7, not represented here because the "Inner" axis maximum was truncated to 8 for better readability). *Marcel* behaves better (probably because user threads are more lightweight), but it still can not achieve a better speedup than 8.16. This is due to the fact that neither NPTL nor *Marcel* takes affinities of threads into account, leading to very frequent remote memory accesses, cache invalidation, etc. We hence used our bubble strategy to distribute the bubble hierarchy corresponding to the nested OpenMP parallelism over the whole machine, and could then achieve better results (up to 10.2 speedup with $16 * 4$ threads). This improvement is due to the fact that the bubble strategy carefully distribute the computation over the machine (on runqueues) in an affinity-aware way (the bubble hierarchy).

It must be noted that for achieving the latter result, the only addition we had to do to the BT-MZ source code is the following line:

```
call marcel_set_load(int(proc_zone_size(myid+1)))
```

that explicitly tells the bubble spread scheduler the load of each zone, so that they can be properly distributed over the machine. Such a clue (which could even be dynamic) is very precious for permitting the runtime environment to make appropriate decisions, and should probably be added as an extension to

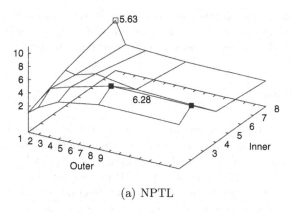

(a) NPTL

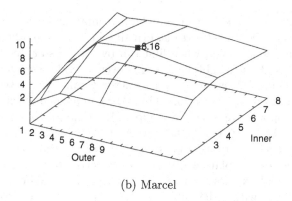

(b) Marcel

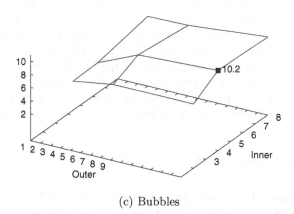

(c) Bubbles

Fig. 5. Nested parallelism

the OpenMP standard. Another way to achieve load balancing would be to create more or less threads according to the zone size [AGMJ04]. This is however a bit more difficult to implement than the mere function call above.

5   Conclusion

In this paper, we discussed the importance of establishing a persistent cooperation between an OpenMP compiler and the underlying runtime system for achieving high performance on nowadays multi-core NUMA machines. We showed how we extended the GNU OpenMP implementation, GOMP, for making use of the flexible *Marcel* thread library and its high-level *bubble* abstraction. This permitted us to implement a scheduling strategy that is suited to OpenMP nested parallelism. The preliminary results show that it improves the achieved speedup a lot.

At this point, we are enhancing our implementation so as to introduce just-in-time allocation for *Marcel* threads, bringing in the notion of "ghost" threads, that would only be allocated when first run by a processor. In the short term, we will keep validating the obtained results over several other OpenMP applications, such as Ondes3D (French Atomic Energy Commission). We will compare the resulting performance with other OpenMP compilers and runtimes. We also intend to develop an extension to the OpenMP standard that will provide programmers with the ability to specify load information in their applications, which the runtime will be able to use to efficiently distribute threads.

In the longer run, we plan to extract the properties of memory affinity at the compiler level, and express them by injecting gathered information into more accurate attributes within the bubble abstraction. These properties may be obtained either thanks to new directives *à la* UPC[4] [CDC+99] or be computed automatically via static analysis [SGDA05]. For instance, this kind of information is helpful for a bubble-spreading scheduler, as we want to determine which bubbles to explode or to decide whether or not it is interesting to apply a *migrate-on-next-touch* mechanism [NLRH06] upon a scheduler decision. All these extensions will rely on a memory management library that attaches information to bubbles according to memory affinity, so that, when migrating bubbles, the runtime system can migrate not only threads but also the corresponding data.

6   Software Availability

Marcel and BubbleSched are available for download within the PM2 distribution at http://runtime.futurs.inria.fr/Runtime/logiciels.html under the GPL license. The MaGOMP port of libgomp will be available soon and may be obtained on demand in the meantime.

---

[4] The UPC `forall` statement adds to the traditional `for` statement a fourth field that describes the affinity under which to execute the loop.

References

[AGMJ04] Ayguade, E., Gonzalez, M., Martorell, X., Jost, G.: Employing Nested OpenMP for the Parallelization of Multi-Zone Computational Fluid Dynamics Applications. In: 18th International Parallel and Distributed Processing Symposium (IPDPS) (2004)

[BS05] Blikberg, R., Sørevik, T.: Load balancing and OpenMP implementation of nested parallelism. Parallel Computing 31(10-12), 984–998 (2005)

[CDC+99] Carlson, W., Draper, J.M., Culler, D.E., Yelick, K., Brooks, E., Warren, K.: Introduction to UPC and Language Specification. Technical Report CCS-TR-99-157, George Mason University (May 1999)

[DGC05] Duran, A., Gonzàles, M., Corbalán, J.: Automatic Thread Distribution for Nested Parallelism in OpenMP. In: 19th ACM International Conference on Supercomputing, Cambridge, MA, USA, June 2005, pp. 121–130 (2005)

[dWJ03] Van der Wijngaart, R.F., Jin, H.: NAS Parallel Benchmarks, Multi-Zone Versions. Technical Report NAS-03-010, NASA Advanced Supercomputing (NAS) Division (2003)

[FLR98] Frigo, M., Leiserson, C.E., Randall, K.H.: The Implementation of the Cilk-5 Multithreaded Language. In: ACM SIGPLAN Conference on Programming Language Design and Implementation (PLDI), Montreal, Canada (June 1998), http://theory.lcs.mit.edu/pub/cilk/cilk5.ps.gz

[gom] GOMP – An OpenMP implementation for GCC, http://gcc.gnu.org/projects/gomp/

[GOM+00] Gonzalez, M., Oliver, J., Martorell, X., Ayguade, E., Labarta, J., Navarro, N.: OpenMP Extensions for Thread Groups and Their Run-Time Support. In: Languages and Compilers for Parallel Computing. Springer, Heidelberg (2000)

[GSS+06] Gao, G.R., Sterling, T., Stevens, R., Hereld, M., Zhu, W.: Hierarchical multithreading: programming model and system software. In: 20th International Parallel and Distributed Processing Symposium (IPDPS) (April 2006)

[MM06] Marathe, J., Mueller, F.: Hardware Profile-guided Automatic Page Placement for ccNUMA Systems. In: Sixth Symposium on Principles and Practice of Parallel Programming (March 2006)

[NLRH06] Nordén, M., Löf, H., Rantakokko, J., Holmgren, S.: Geographical Locality and Dynamic Data Migration for OpenMP Implementations of Adaptive PDE Solvers. In: Second International Workshop on OpenMP (IWOMP 2006), Reims, France (2006)

[SGDA05] Shen, X., Gao, Y., Ding, C., Archambault, R.: Lightweight Reference Affinity Analysis. In: 19th ACM International Conference on Supercomputing, Cambridge, MA, USA, June 2005, pp. 131–140 (2005)

[Thi05] Thibault, S.: A Flexible Thread Scheduler for Hierarchical Multiprocessor Machines. In: Second International Workshop on Operating Systems, Programming Environments and Management Tools for High-Performance Computing on Clusters (COSET-2), Cambridge / USA, 06 2005. ICS / ACM / IRISA

[TTSY00] Tanaka, Y., Taura, K., Sato, M., Yonezawa, A.: Performance Evaluation of OpenMP Applications with Nested Parallelism. In: Languages, Compilers, and Run-Time Systems for Scalable Computers, pp. 100–112 (2000)
[Zha06]   Zhang, G.: Extending the OpenMP standard for thread mapping and grouping. In: Second International Workshop on OpenMP (IWOMP 2006), Reims, France (2006)

Problems, Workarounds and Possible Solutions Implementing the Singleton Pattern with C++ and OpenMP

Michael Suess and Claudia Leopold

University of Kassel, Research Group Programming Languages / Methodologies,
Wilhelmshöher Allee 73, D-34121 Kassel, Germany
{msuess,leopold}@uni-kassel.de

Abstract. Programs written in C++ and OpenMP are still relatively rare. This paper shows some problems with the current state of the OpenMP specification regarding C++. We illustrate the problems with various implementations of the singleton pattern, measure their performance and describe workarounds and possible changes to the specification. The singletons are available in the AthenaMP open-source project.

1 Introduction

Programming with OpenMP and C++ is still challenging today. With present compilers, even basic C++ features such as exceptions often do not work. To provide more test cases and experiment with advanced C++ features in combination with OpenMP, the AthenaMP [1] open source project was created. This work is part of AthenaMP.

One of the most widely known patterns among programmers is the *singleton*. It has been thoroughly analyzed and implemented. This paper deploys thread-safe singletons as an example to show how OpenMP and C++ can be used together. Several implementations are described in depth, which is our first contribution. The second contribution is pointing out problems with the use of OpenMP in combination with C++. We provide workarounds for these problems and suggest changes to the OpenMP specification to solve them fully.

The paper is organized as follows: In Section 2, the singleton pattern is introduced, along with a simple, non-threaded implementation in C++. Section 3 describes various thread-safe implementations, highlights problems with OpenMP and explains possible workarounds, while their performance is benchmarked in Section 4. Some implementations require changes to the OpenMP specification, which are collected in Section 5. The work of this paper is embedded in the AthenaMP open-source project described in Section 6. The paper finishes with related work in Section 7, and a summary in Section 8.

2 The Singleton Pattern

The most famous description of the intent of the singleton pattern is from Gamma et al. [2]:

B. Chapman et al. (Eds.): IWOMP 2007, LNCS 4935, pp. 173–184, 2008.

Ensure a class only has one instance, and provide a global point of access to it.   (Gamma et al. [2])

Singletons are useful to ensure that only one object of a certain type exists throughout the program. Possible applications are a printer spooler or the state of the world in a computer game. While singletons provide a convenient way to access a resource throughout the program, one needs to keep in mind that they are not much more than glorified global variables, and just like them need to be used with care (or not at all, if possible). Yegge explains this in great detail in one of his weblog posts [3].

While singletons can be implemented in C++ using inheritance, we have decided to implement them using wrappers and templates, as described by Schmidt et al. [4]. These authors call their classes adapters, but we stick to the more general name *singleton wrapper* instead, to avoid confusion with the well-known adapter pattern (which is different from what we are doing). We provide wrapper classes that can be instantiated with any class to get the singleton functionality. For example, to treat the class `my_class` as a singleton and call the method `my_method` on it, one needs to do the following:

```
singleton_wrapper <my_class >:: instance (). my_method ();
```

Provided that all accesses to the class `my_class` are carried out in this way, then and only then the class is a singleton. The code of class `my_class` does not need to be changed in any way, but all of the singleton functionality is provided by the wrappers. This is the biggest advantage of this implementation over an inheritance-based approach. Typical requirements for an implementation state that the singleton:

— must not require prior initialization, because forgetting to do so is a frequent mistake by programmers
— must be initialized only when it is needed (lazy initialization)
— must return the same instance of the protected object, regardless of whether it is called from within or outside a parallel region

A non-threadsafe version of the `instance` method of a singleton wrapper is shown in Figure 1. Here, `instance_` is a private static member variable, omitted for brevity in all of our examples. This implementation, of course, has problems in a multi-threaded environment, as the `instance_` variable is a shared resource and needs to be protected from concurrent access. Ways to deal with this problem are shown in the next section.

3   Thread-Safe Singleton Implementation Variants

A safe and simple version of a multi-threaded singleton wrapper is shown in Figure 2. It uses a named critical region to protect access to the singleton instance. While this solution solves the general problem of protecting access to

```
template <class Type>
class singleton_wrapper {
static Type& instance ()
{
  if (instance_ == 0) {
    instance_ = new Type;
  }
  return *instance_;
}
};
```

Fig. 1. The `instance` method of a sequential singleton wrapper

```
template <class Type>
class singleton_wrapper {
static Type& instance ()
{
  #pragma omp critical (ATHENAMP_1)
  {
    if (instance_ == 0) {
      instance_ = new Type;
    }
  }
  return *instance_;
}
};
```

Fig. 2. The `instance` method of a simple, thread-safe singleton wrapper

the singleton class, it has two major drawbacks, both of which we are going to solve later: First, it uses the same named critical construct to protect all classes. If e. g. a singleton for a printer spooler and a singleton for the world state is needed in the same program, access to these classes goes through the same critical region and therefore these accesses can not happen concurrently. This restriction is addressed in Section 3.1. The second problem is that each access to the singleton has to pay the cost of the critical region, although technically it is safe to work with the singleton after it has been properly initialized and published to all threads. An obvious (but incorrect) attempt to solve this problem is shown in Section 3.2, while Sects. 3.3 and 3.4 solve the problem, but have other restrictions.

3.1   The Safe Version Using One Lock Per Protected Object

In the last section, we have shown a thread-safe singleton that used one critical region to protect accesses to all singletons in the program. This introduces unnecessary serialization, as a different critical region per singleton is sufficient. We will show four different ways to solve the problem. The first two require changes in the OpenMP specification to work but are quite simple from a programmer's point of view, the third can be done today but requires a lot of code. The fourth solution requires a helper-class and has problems with some compilers.

Attempt 1: Extending Critical: Our first attempt at solving the problem is shown in Figure 3. The idea is to give each critical region a unique name by using the template parameter of the singleton wrapper. Unfortunately, this idea does not work, because the compilers treat the name of the critical region as a string and perform no name substitution on it. While it would be theoretically feasible to change this in compilers, because template instantiation happens at compile-time, it would still require a change in the OpenMP specification, and we suspect the demand for that feature to be small. For this reason, we are not covering this attempt any further in Section 5.

```
// this code works, but does
// not solve the problem!
template <class Type>
class singleton_wrapper {
static Type& instance ()
{
  #pragma omp critical(Type)
  {
    if (instance_ == 0) {
      instance_ = new Type;
    }
  }
  return *instance_;
}
};
```

```
// this code does not work!
template <class Type>
class singleton_wrapper {
static Type& instance ()
{
  static omp_lock_t my_lock
    = OMP_LOCK_INIT;
  omp_set_lock (&my_lock);
  if (instance_ == 0) {
    instance_ = new Type;
  }
  omp_unset_lock (&my_lock);
  return *instance_;
}
};
```

Fig. 3. instance method with multiple critical regions - Attempt 1

Fig. 4. instance method with multiple critical regions - Attempt 2

Attempt 2: Static Lock initialized with OMP_LOCK_INIT: Our second attempt to solve the problem uses OpenMP locks. The code employs a static lock to protect access to the shared instance variable. Since each instance of the template function is technically a different function, each instance gets its own lock as well, and therefore each singleton is protected by a different critical region.

The big problem with this approach is to find a way to initialize the lock properly. There is only one way to initialize a lock in OpenMP – by calling omp_init_lock. This must only be done once, but OpenMP does not provide a way to carry out a piece of code only once (a solution to this shortcoming is presented later in this section). One of our requirements for the singleton is that it must not need initialization beforehand, therefore we are stuck.

A solution to the problem is shown in Figure 4 and uses static variable initialization to work around the problem, by initializing the lock with the constant OMP_LOCK_INIT. This is adapted from POSIX Threads, where a mutex can be initialized with PTHREAD_MUTEX_INITIALIZER. In a thread-safe environment (which OpenMP guarantees for the base language), the runtime system should make sure this initialization is carried out only once, and all the compilers we have tested this with actually do so. Of course, OMP_LOCK_INIT is not in the OpenMP specification, but we believe it would be a worthy addition to solve this and similar problems, not only related to singletons. Therefore, although quite an elegant solution, this attempt does not work with OpenMP today as well.

Attempt 3: Doing It Once: As shown in the previous paragraph, method omp_init_lock needs to be called only once, but OpenMP provides no facilities to achieve that. It is possible to code this functionality in the program itself, though, as one of the authors has described in his weblog [5]. The necessary methods to achieve this are available and described in the AthenaMP library [1] as well as the above-mentioned weblog entry. Here, we restrict ourselves to showing how the functionality can be used.

Figure 5 depicts the instance method with once functionality. Line 19 has the actual call to the once template-function defined in AthenaMP. The function

```
 1   struct init_func {
 2     omp_lock_t *lock;
 3     void operator() ()
 4     {
 5       omp_init_lock (lock);
 6     }
 7   };
 8
 9   template <class Type>
10   class singleton_wrapper {
11   static Type& instance ()
12   {
13     static omp_lock_t my_lock;
14     static once_flag flag
15       = ATHENAMP_ONCE_INIT;
16     init_func my_func;
17     my_func.lock = &my_lock;
18
19     once (my_func, flag);
20
21     omp_set_lock (&my_lock);
22     if (instance_ == 0) {
23       instance_ = new Type;
24     }
25     omp_unset_lock (&my_lock);
26     return *instance_;
27   }
28   };
```

```
template <class Type>
class singleton_wrapper {
static Type& instance ()
{
    static omp_lock_ad my_lock;
    my_lock.set ();
    if (instance_ == 0) {
      instance_ = new Type;
    }
    my_lock.unset ();
    return *instance_;
}
};
```

Fig. 5. instance method with multiple critical regions - Attempt 3

Fig. 6. instance method with multiple critical regions - Attempt 4

takes two parameters, the first one being a functor that includes what needs to happen only once (also shown at the top of Figure 5). The second parameter is a flag of the type `once_flag` that is an implementation detail to be handled by the user.

This attempt to solve our problem works. Nevertheless, it needs a lot of code, especially if you also count the code in the `once` method. Moreover, it relies on static variable initialization being thread-safe, as our last attempt. For these reasons, we are going to solve the problem one last time.

Attempt 4: Lock Adapters to the Rescue: Figure 6 shows our last attempt at protecting each singleton with its own critical region. It is substantially smaller than all previous versions and should also work with OpenMP today. It uses lock adapters (called `omp_lock_ad`) that are part of AthenaMP and are first introduced in a different publication by the same authors [6].

Lock adapters are small classes that encapsulate the locking functionality provided by the parallel programming system. In AthenaMP, two of these classes are included, `omp_lock_ad` for simple OpenMP locks and `omp_nest_lock_ad` for nested OpenMP locks. Lock adapters for different parallel programming systems are trivial to implement. The initialization code for the locks is in the constructors of the adapter classes, therefore their initialization happens automatically. Destruction of the encapsulated lock is done in the destructor of the adapter. A common mistake in OpenMP (forgetting to initialize a lock before using it) is therefore not possible when using lock adapters. For more details about the lock adapters please refer to another publication by the same authors [6].

Because lock initialization happens automatically when an instance of the class omp_lock_ad is created, our problem with the initialization having to be carried out only once disappears. Or at least: it should disappear. Unfortunately, some compilers we have tested this with call the constructor of the lock adapter more than once in this setting, although it is a shared object. The next paragraph explains this more thoroughly.

The difference between static variable construction and static variable initialization becomes important here, because static variable initialization works correctly in a multi-threaded setting with OpenMP on all compilers we have tested, but static variable construction has problems on some compilers. Static variable initialization basically means declaring a variable of a primitive data type and initializing it at the same time:

```
static int my_int = 5;
```

This makes my_int a shared variable and as in the sequential case, it's initialization is carried out once. This works on all compilers we have tested. Static variable construction on the other hand looks like this:

```
static my_class_t my_class;
```

Since my_class_t is a user-defined class, it's constructor is called at this point in the program. Since the variable is static, it is a shared object and therefore the constructor should only be called once. This does not happen on all compilers we have tested (although we believe this to be a bug in them, since OpenMP guarantees thread-safety of the base language) and is therefore to be used with care. A trivial test program to find out if your compiler correctly supports static variable construction with OpenMP is available from the authors on request.

This solution is the most elegant and also the shortest one to provide each singleton with its own critical region. Nevertheless, our second problem is still there: each access to the singleton has to go through a critical region, even though only the creation of the singleton needs to be protected. We are going to concentrate on this problem in the next section.

3.2   Double-Checked Locking and Why It Fails

Even in some textbooks, the *double-checked locking* pattern is recommended to solve the problem of having to go through a critical region to access a singleton [4]. Our instance method with this pattern is shown in Figure 7.

Unfortunately, the pattern has multiple problems, among them possible instruction reorderings by the compiler and missing memory barriers, as explained by Meyers and Alexandrescu [7], as well as problems with the OpenMP memory model, as explained by de Supinski on the OpenMP mailing list [8]. Since the pattern may and will fail in most subtle ways, it should not be employed.

```
// this code is wrong!!
template <class Type>
class singleton_wrapper {
static Type& instance ()
{
  #pragma omp flush(instance_)
  if (instance_ == 0) {
    #pragma omp critical(A_2)
    {
      if (instance_ == 0)
        instance_ = new Type;
    }
  }
  return instance_;
}
};
```

```
template <class Type>
class singleton_wrapper {
static Type* cache_;
static Type& instance ()
{
  #pragma omp threadprivate(cache_)
  if (cache_ == 0) {
    cache_ =
      &singleton<Type>::instance ();
  }
  return *cache_;
}
};
```

Fig. 7. instance method with double-checked locking

Fig. 8. instance method using caching

3.3 Using a Singleton Cache

Meyers and Alexandrescu [7] also suggest caching a pointer to the singleton in each thread, to avoid hitting the critical region every time the instance method is called. This can of course be done by the user, but we wanted to know if it was possible to extend our singleton wrapper to do this automatically. We therefore came up with the implementation shown in Figure 8.

The implementation solves the problem, as the critical region for each singleton is only entered once. It relies on declaring a static member variable threadprivate. Unfortunately, the OpenMP specification does not allow to privatize static member variables in this way. In our opinion, this is an important omission not restricted to singletons, and therefore this point is raised again later in this paper, when we put together all enhancement proposals for the OpenMP specification in Section 5.

There is a workaround, however, which was suggested by Meyers in his landmark publication Effective C++ [9] as a solution to a different problem. Instead of making the cache a static member variable (that cannot be declared threadprivate), it can be declared as a static local variable in the instance method. Declaring local variables threadprivate is allowed by the specification, and this solves the problem without any further disadvantages.

The whole solution unfortunately has some problems. It relies on threadprivate data declared with the threadprivate directive, but in OpenMP these have some restrictions. In a nutshell, these data become undefined as soon as nested parallelism is employed or as soon as the number of threads changes over the course of multiple parallel regions (see the OpenMP specification for details [10]). There is no way to work around these limitations for this solution, therefore the user has to be made aware of them by carefully documenting the restrictions. Fortunately, there is a different way to achieve the desired effect and it is explained in the next section.

```
template <class Type>
class singleton_wrapper {
static Type& instance ()
{
    static Type instance;
    return instance;
}
};
```

Fig. 9. `instance` method using a Meyers singleton

3.4   The Meyers Singleton

One of the most well-known singleton implementations today is the so-called Meyers Singleton that is described in Effective C++ [9]. It is quite elegant, small, and shown in Figure 9.

Meyers himself warns about using his implementation in a multi-threaded setting, because it relies on static variable construction being thread-safe. Of course, Meyers did not write about OpenMP. OpenMP guarantees thread-safety of the base language in the specification, and this should cover proper construction of static variables in a multi-threaded setting as well. Unfortunately, as described in Section 3.1 in detail, our tests show that some OpenMP-aware compilers still do have problems with this. Some of them were even calling the constructor of the same singleton twice, which should never happen in C++. Therefore, although it is the smallest and most elegant solution to the problem, it cannot be recommended for everyone at this point in time.

4   Performance

This section discusses the performance of the proposed singleton wrapper implementations. We have setup a very simple test to access the performance of our solution. It is shown in Figure 10.

```
int counter = 0;
double fcounter = 0.0;

double start = omp_get_wtime ();

singleton_wrapper<int >::instance() = 1;
singleton_wrapper<double>::instance() = 1.0;
#pragma omp parallel reduction (+:counter) reduction (+:fcounter)
{
    for (int i=0; i<numtries; ++i) {
        counter += singleton_wrapper<int >::instance();
        fcounter += singleton_wrapper<double>::instance();
    }
}

double end = omp_get_wtime ();
```

Fig. 10. The code used to benchmark our singletons

Table 1. Measured benchmark timings in seconds, best of three runs, 10.000.000 Singleton accesses per thread - only the entries printed in bold are correct and safe!

Test Environment	one_crit	multi_crit	local_cache	meyers	dcl
AMD, Intel Comp., 4 Thr.	**32.8**	**18.6**	**1.38**	**0.04**	1.46
Sun, Sun Comp., 8 Thr.	**176**	**182**	n.a.	0.14	0.62
IBM, IBM Comp., 8 Thr.	**72.7**	**71.7**	**0.34**	0.01	0.85

Two different singletons are used in the example, one is an integer and one is a double. The protected singletons will usually be classes, of course, but for our simple performance measurements primitive data-types will do. The singletons are initialized prior to the parallel region. Inside the parallel region, they are read only and their result is added up and tested outside the parallel region for correctness (not shown in the figure). The results of our tests are shown in Table 1.

Here is a short summary of the table headings:

- **one_crit:** a singleton wrapper using the same critical region for all protected singletons (see Figure 2)
- **multi_crit:** a singleton wrapper using a different critical region for all protected singletons with a lock adapter (see Figure 6)
- **local_cache:** a singleton wrapper that caches a pointer to the singleton in threadprivate memory (see Figure 8)
- **meyers:** a singleton wrapper built after the Meyers Singleton (see Figure 9) - the numbers are only representative on the Intel Compiler, because the other two do not construct static classes correctly
- **dcl:** for comparison, we have also included the double-checked locking version (see Figure 7), although it is not safe to use!

As can be clearly seen by these numbers, the Meyers Singleton is to be preferred on all architectures, as it is the fastest by several orders of magnitude. We don't have any numbers for the version using a threadprivate cache on the Sun Compiler, as it was not able to translate our code. We believe this to be a bug in the compiler.

These results leave us with a disappointing situation: we have isolated a best solution (Meyers Singleton), the solution should be legal judging from the OpenMP specification, yet some compilers don't implement it correctly. While we cannot fix the compilers, what we can do at this point is provide a short test program that shows which compilers behave correctly and which do not. It is available from the authors on request.

5   OpenMP Enhancement Proposals for C++

In this section, we suggest enhancements to the OpenMP specification that we found useful during the course of our work on AthenaMP. Most of them have been sketched earlier in this publication already and are summarized here to have them all in one place. All of them are useful beyond our simple singleton implementations in our opinion.

Lock Initialization with OMP_LOCK_INIT: We have shown in Section 3.1 that it would make sense to initialize OpenMP locks not only by using the omp_init_lock function, but also with a macro, e.g. OMP_LOCK_INIT. This is adapted from POSIX Threads, where PTHREAD_MUTEX_INITIALIZER can be used to initialize a lock. If this idea was accepted into the standard, it would be possible to initialize static locks using static variable initialization, a facility that is guaranteed to happen only once in OpenMP.

Declaring Static Member Variables threadprivate: In Section 3.3 we have shown that there is benefit in allowing static member variables to be declared threadprivate. This would be a very simple change in the specification and has already been implemented in the Intel Compiler.

Flushing Reference Variables not Allowed: Although it is not explicitly forbidden in the specification to flush reference variables, most compilers do not allow it either. A very common use case for reference variables is to pass parameters to functions by reference, either because they need to be changed inside the function or because the object is large and copying it would include a performance penalty. This is a recommended practice in many textbooks about C++. We have hit this problem when implementing the once functionality touched in Section 3.1, where we pass the second parameter to the once function by reference to const and would like to flush it inside the function.

The specification only allows to flush pointers and not pointees. This is a problem in our case, because reference variables are most likely implemented as pointers. The OpenMP specification should therefore explicitly allow the special case of flushing reference variables to be more conforming to recommended C++ practices.

6   AthenaMP

The functionality presented in this paper is part of the AthenaMP open source project [1]. The main goal of this project is to provide implementations for a set of concurrent patterns using OpenMP and C++. It includes both low-level patterns like advanced locks, and higher-level ones. The patterns demonstrate solutions to parallel programming problems as a reference for programmers, and additionally can be used directly as generic components. The code is also useful for compiler vendors testing their OpenMP implementations against more involved C++ code, an area where many compilers today still have difficulties. A more extensive project description is provided by one of the authors in his weblog [11].

7   Related Work

Lots of work has been put into correctly implementing the singleton pattern. The most famous resource on the topic is the book by Gamma et al. [2]. The idea for our singleton wrapper is described by Schmidt et al. [4], along with double-checked locking that was later proved to be an anti-pattern by Meyers and Alexandrescu [7]. The idea for the singleton cache is also from the latter source. More involved descriptions of singletons with different properties in C++ are given by Alexandrescu [12]. Yegge describes most clearly why singletons should be used with care [3].

8 Summary and Contributions

This work features two main contributions. First, we have described and evaluated different implementation possibilities for a thread-safe singleton with C++ and OpenMP (Section 3), a work that has to our knowledge never been attempted using these systems before. Second, we have highlighted some problems encountered with the present OpenMP specification and C++. Among them are the inability to initialize static locks with a macro, the inability to declare static member variables threadprivate, and the problem of not being allowed to flush reference variables.

Acknowledgments

We are grateful to Björn Knafla for proofreading the paper and for guiding us through the dark corners of C++ when needed. We thank the University Computing Centers at the RWTH Aachen, TU Darmstadt and University of Kassel for providing the computing facilities used to carry out our unit tests on different compilers and architectures.

References

1. Suess, M.: AthenaMP (2007), http://www.athenamp.org/
2. Gamma, E., Helm, R., Johnson, R., Vlissides, J.: Design Patterns: Elements of Reusable Object-Oriented Software. Addison-Wesley Professional, Reading (1995)
3. Yegge, S.: Singleton Considered Stupid (2004),
   http://steve.yegge.googlepages.com/singleton-considered-stupid
4. Schmidt, D.C., Stal, M., Rohnert, H., Buschmann, F.: Pattern-Oriented Software Architecture. Networked and Concurrent Objects, vol. 2. John Wiley and Sons, Chichester (2000)
5. Suess, M.: How to do it ONCE in OpenMP (2006), http://www.thinking-parallel.com/2006/09/20/how-to-do-it-once-in-openmp/
6. Suess, M., Leopold, C.: Generic Locking and Deadlock-Prevention with C++. In: Proceedings of Parallel Computing (ParCo 2007) (to appear, 2007)
7. Meyers, S., Alexandrescu, A.: C++ and the Perils of Double-Checked Locking - Part 1. Dr. Dobb's Journal, 46–49 (2004)
8. de Supinski, B.R.: [Omp] A memory model question (2006),
   http://openmp.org/pipermail/omp/2006/000479.html
9. Meyers, S.: Effective C++: 55 Specific Ways to Improve Your Programs and Designs, 3rd edn. Addison-Wesley Professional, Reading (2005)

10. OpenMP Architecture Review Board: OpenMP Specifications (2005),
    http://www.openmp.org/specs
11. Suess, M.: A Vision for an OpenMP Pattern Library in C++ (2006), http://www.
    thinkingparallel.com/2006/11/03/a-vision-for-an-openmp-pattern-library
    -in-c/
12. Alexandrescu, A.: Modern C++ Design: Generic Programming and Design Pat-
    terns Applied. Addison-Wesley Professional, Reading (2001)

Web Service Call Parallelization Using OpenMP

Sébastien Salva[1], Clément Delamare[2], and Cédric Bastoul[3]

[1] LIMOS, Université d'Auvergne Clermont 1
Campus des Cézeaux
F-63173 Aubière, France
salva@iut.u-clermont1.fr
[2] DGI, Direction Générale des impôts, France
86 allée de Bercy, F-75012 Paris, France
clement.delamare@dgi.finances.gouv.fr
[3] LRI, ALCHEMY, Université Paris-Sud
Bâtiment 490, F-91405 Orsay Cedex, France
cedric.bastoul@lri.fr

Abstract. Since early works, web servers have been designed as parallel applications to process many requests at the same time. While web service based applications are performing more and more, larger and larger transactions, this parallelization culture still not reached the client side. Business to Business (B2B) applications are becoming intensive users of web transactions through Service Oriented Architecture Standard. As multicore systems are now widely available, parallelization seems the right way to fit the need of such applications. In this paper, we describe an API based on OpenMP for transparently parallelizing web service calls, i.e. the serialization, deserialization and connection processes. Our API is mainly based on a software pipeline which splits web service calls into several tasks. Both synchronous and asynchronous modes can be used with this API to call web services. We present experimental evidence demonstrating the ability of our API to achieve high level performance.

1 Introduction

SOA (Service Oriented Architecture) is emerging as the standard paradigm to develop business applications over Internet, like Business to Business (B2B) and Business to Consumer (B2C) applications which involve the transaction of goods or services. Such applications are mainly based on web service interactions. Web services represent objects whose methods can be called through Internet. They generally accept parameters (objects) and generate a response. Both parameters and responses are serialized/deserialized by using an XML based protocol, named SOAP (Service Oriented Architecture Protocol).

Business applications, like banking systems, require a constantly growing number of large data transactions. Those data flood capacity as the throughput is increasing by ten times nowadays. As a consequence, such applications require more and more performance, especially for the costly data serialization/deserialization processes. For B2B applications which perform a lot of web service

B. Chapman et al. (Eds.): IWOMP 2007, LNCS 4935, pp. 185–194, 2008.

transactions, the use of monocore systems is becoming a limiting factor. Because of the inherently parallel nature of web service calls, the trend is to benefit from multicore systems to improve server capacity, as Sun T1 (Niagara) architecture shows. However, a trivial parallelization scheme does not reach the maximal service performance because of the limited number of cores and a suboptimal use of resources when server threads stall, waiting for other server responses.

In this paper, we present a solution using OpenMP to efficiently parallelize web service based applications on multicore systems. We describe an API for transparently parallelizing web service calls on multicore systems. This API is easy to use: the developer gives the web services to call and its code to manage web service responses (they can be stored or analyzed to produce other calls).

Our API is mainly based on a software pipeline whose stages split web service calls into independent tasks and parallelize their execution. We show that this solution performs better than simple parallelization schemes when using both synchronous web service which responds in some milliseconds, and asynchronous ones whose responses may be received from some seconds to some days later. Because Java is widely used for coding web service based applications, our API is written using this language. We have also used the OpenMP Jomp API [1] for the parallelization. OpenMP brings many advantages, especially for the software pipeline generation and for the parallelization of the serialization/deserialization processes, which are mainly based on loops.

The remainder of this paper is structured as follow: Section 2 provides an overview of the web service paradigm and of a web service based project, Copernic [2]. Then, we present some related works. Section 3 describes our API mainly based on a software pipeline. Section 4 shows some experimentations. Finally, Section 5 gives some perspectives and conclusions.

2   Web Services Overview and Related Work

Web services are "self contained, self-describing modular applications that can be published, located, and invoked across the web" [3]. The Web Service framework, which is a platform independent standard, is composed of three major axes:

- The *service description* which models how to invoke the service by defining its interface and its parameters. This description, called WSDL (Web Services Description Language) file [4], gathers several parts describing types, accessible methods, query and response message structures, binding and transport options.
- The *definition and the construction of XML messages*, based on the Simple Object Access Protocol (SOAP). SOAP aims to invoke service operations (object methods) on the Internet and to serialize/deserialize data. When SOAP is used over the HTTP layer, the web service is said to be called in *synchronous mode*. When it is used over the SMTP layer, it is said to be called in *asynchronous mode*.
- The *discovery of the service*. Web services are gathered into UDDI (Universal Description, Discovery Integration) registries. These registries can be

consulted manually or automatically using dedicated APIs to find specific web services. The WSDL files are grouped into these registries.

Web services allow to externalize functional code of companies in a standardized way. Web Services are mainly used for B2B or B2C transactions, to exchange data and externalize functional code.

An example of advanced SOA based project is Copernic [2]. Copernic is a modernization program that will recast and upgrade the entire fiscal information system in France. At the moment, it covers 70 individual projects, managed by the Direction Générale des Impôts and the Direction Générale de la Comptabilité Publique. Its objective is to enable the French tax administration to offer new, citizen-centered services and to boost the efficiency of its internal processes. A lot of data is managed with Copernic, like the PERS reference which models a citizen or the TOPAD reference which models all the French roads with postal addresses. Such data are accessed by web services to modify or consult them. Copernic produces large web service transactions, especially with specific tasks, like the edition of the 33 millions income tax returns which uses many references (PERS, TOPAD and many others ones).

Experiences, to parallelize web servers using OpenMP, have been gained by Balart et al. [5], but few solutions have been proposed to parallelize web service calls even in the general case. Kut and Birant suggested an approach which uses some threads to call several web services in the same time [6]. We show, in Section 3.1 that this method should not be used when the response receipt delay is long (especially with asynchronous web services). Wurz and Schuldt consider another approach for web service parallelization: if the original request is composed of several independent sub-requests, this one is split and the sub-requests are executed in parallel [7]. This solution offers good results, however most of web services use atomic requests (i.e., they are not composed of sub requests).

3   Web Service Call Parallelization

SOA based applications (for e-commerce, B2B or B2C applications) perform large transactions with web services. There are several advantages to parallelize such services: from being able to call several other services at the same time, up to reduce the time necessary to serialize/deserialize data.

Here, we focus on the client side, i.e., applications that call successively web services which may produce themselves other web service calls. In this section, we propose and describe an API which aims at easily and efficiently constructing such applications. For a developer, this API is mainly based on a class named ParallelClient, which is used to set the number of working threads and to add web service calls. The developer has only to focus on the implementation of the web service responses management: data received from web services can be either stored or analyzed. The rest of the client application, i.e., the data serialization/deserialization processes and the web service calls, is made in parallel transparently thanks to the API.

Web service based applications are typically written in C# or in Java. To construct a parallel client API, we chose Java and the Jomp OpenMP API [1]. Jomp is a compiler translating Java source code with OpenMP directives to another code using Java threads to implement parallelism.

In the following, we consider that a complete web service call is composed of the following steps:

1. the S step: a data serialization into a SOAP-XML document,
2. the C step: a web service connection, that is a TCP connection with an Internet service provider to send the serialized data and to receive a SOAP-XML response,
3. the D step: a response deserialization whose memory size may be important (100MB or even more)
4. the P step: obtained data are stored in a persistent infrastructure like a database, or may be analyzed.

3.1   A Naive Solution: Using the Task Pool Paradigm

Because of the inherently parallel nature of web service calls, a basic solution would be to use a task pool where one task represents one web service call (including all the steps previously defined). Therefore, with this solution, if n threads are available, n calls can be easily done in parallel, assuming n cores are available. The task pool algorithm is described in Figure 1.

```
 1  pool is a FIFO containing tasks;
 2  setNumThreads(n);
 3
 4  // omp parallel shared(pool) private(....){
 5  // omp critical {
 6  task=read(pool);}
 7  while(task != NULL){
 8      SOAPdata=Serialize(task.data);
 9      Response=WebService.call(SOAPdata, endpoint);
10      Object [] obj=Deserialize(Response);
11      Persistency(obj);
12      // omp critical {
13      task=read(pool);}
14      }
15  }
```

Fig. 1. The task pool pseudocode

Figure 2 illustrates a task pool execution with only two threads. After each web service connection, each thread must wait for the web service data receipt before deserializing it. It follows, for n threads after n web service connections, the application may be blocked. When web services are called in synchronous

Fig. 2. Two threads executing two web service calls

mode, the application is blocked during some seconds. However, in asynchronous mode, the client application may be blocked during some days. Furthermore if the target architecture has less than n cores (or n processors for symmetric multiprocessor architectures) this policy will result in thread interleaving, with a potentially high cost induced by thread management and context switching. Hence this solution does not appear as the best one to parallelize web service calls.

3.2   A Better Resource Management Using the Pipeline Paradigm

The web service call steps are executed successively and are completely independent. As a result the pipeline paradigm can be used to parallelize web service calls. We use a pipeline of four stages, one for each step described previously: a serialization stage (S), a web service call stage (C), a deserialization stage (D) and a stage for the persistency (P). This scheme is depicted in Figure 3(a).

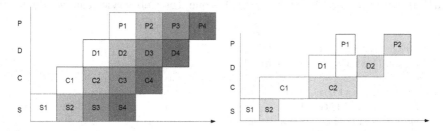

(a) A software pipeline for web service calls

(b) The software pipeline blocked by web service connections

Fig. 3.

This solution requires at least four cores, each of them being dedicated to a thread processing a given step. This solution avoids thread interleaving and achieves a better use of resources since the system overlaps service call stalling time (while waiting for a response) with processing.

However, such a pipeline architecture also has some drawbacks:

- The web service connection and the waiting of its response is usually a blocking operation. As the response receipt delay may be long (until more than one day in asynchronous mode), the pipeline may be blocked as well. Therefore, the two last stages may stay empty until a response is received.

In Figure 3(b), we illustrate two web service calls: the C stage must receive the response of the first web service before calling the second one. As a consequence, the pipeline is blocked. In this case, there are a few advantages to parallelize this step.

– The serialization and deserialization processes may require long time executions. For instance, a deserialization of 100MB of SOAP-XML data may require more than 30 minutes with a recent monocore processor. As a consequence, the load balancing between the pipeline stages may be catastrophic if there is a lot of data to serialize/deserialize. So these two steps must be also parallelized to reduce the execution time of the serialization/deserialization stages.

Despite these drawbacks, the present solution appears to be better than the trivial one. Indeed the S stage can always work until there are web services to call, since it is never blocked.

3.3 An Optimized Pipeline for Web Service Calls

A refinement of the previous four stage pipeline solution is to extend the first three stages. We do not develop the Persistency stage here because this one depends mainly on the data storage.

The Call Stage. For the web service call stage, we use the ProviderConnection architecture proposed by SUN [8]. In this one, the web service connection is not

```
1   FIFO fifo1 , fifo2 , fifo3 ;
2   //omp parallel sections {
3
4   //serialization code
5   task= read(fifo1);
6   task2.serializeddata=serialize(task.data);
7   write(task2,fifo2);
8   }
9   //omp section {
10  //web service connection code
11  ...
12  }
13  //omp section {
14  //deserialization code
15  ...
16  }
17  //omp section {
18  //data storage code or data analysis
19  ...
20  }
21  }
```

Fig. 4. The pipeline pseudocode

made directly by the client application but by using a service (daemon) which can be localized in the same computer than the Client or in a gateway. In our case, this service is composed of two servlets, the first one is used for calling the web services, and the other one receives the web service responses. So, with such an architecture, we perform non blocking web service calls.

The call stage, illustrated in Figure 5, is finally composed of a task pool where a task is an object {call identifier, URL, method, Message} which models a web service call. An available thread executes a task (1). And it calls the web service by using the Send servlet (2). The Send servlet makes a connection to the web service and sends SOAP-XML data (3). The Response servlet receives the response (4). Each response is given to the Deserialization stage (5).

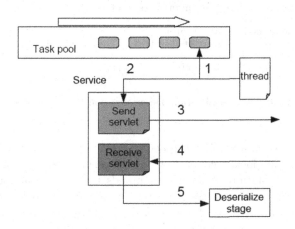

Fig. 5. The call stage architecture

By using such an architecture, the stages are never blocked. The S stage can work until there is web services to call. And the C stage can now establish all the connections. When a response is received, this one is deserialized by the D stage and stored by the P one. Each web service call is numbered by an unique identifier, thus responses may be received in any order. In the synchronous mode, all the stage works. However, in asynchronous mode, if the receipt delay is long, some stages may become idle. So, in this case, the load balancing is not good. We give two approaches to solve this problem in the perspectives.

The Serialization/Deserialization stages. Serialization and deserialization stages are both parallelized using a team of threads. As the data (objects) to serialize are independent, we use one thread team to serialize in parallel all the objects, then the final XML-SOAP document is generated. Similarly, all the elements of a SOAP-XML document can be extracted and then deserialized in parallel by another team of threads. The algorithms of these stages are given in Figure 6:

We show in the following section that the performance of the optimized pipeline is better than the task pool solution. And more precisely, the longer

```
1    Serialization Stage
2    Input: List of nb objects
3    Output: SOAP_XML_document document
4    A list Lfinal of nb empty elements is defined
5    // omp parallel for private(element,i) shared (List,Lfinal)
6    for(i=0,i<nb;i++){
7         Read List[i];
8         (XML element)Lfinal[i]=Serialize L[i];}
9    for(i=0;i<nb;i++)
10        document.add(Lfinal[i]);
11
12   Deserialization Stage
13   Input: SOAP_XML_document document
14   Output: List_object of nb objects
15   nb=numbers of elements(objects) in the SOAP document
16   A list L of nb empty elements is defined
17   for(i=0;i<nb;i++)
18        L[i]=document.get_element(i);
19   //omp parallel for private(object,i) shared (List,L)
20   for(i=0,i<nb;i++){
21        List_object[i]=Deserialize L[i];}
```

Fig. 6. The Serialization/Deserialization stage pseudocode

is the web service data receipt, the better are the performances. This is a consequence of the unblocked connection to web services that the pipeline brings. For a task pool solution, if we use n threads for $2n$ transactions, the first n transactions are performed, then the next ones are done too when the first n transactions are finished. With the optimized pipeline solution, all the transactions are executed in succession and responses are given to the deserialization stage in succession too. By considering only the web service connection time, if this one takes on average 5 minutes, the task pool solution needs at least 10 minutes to finish $2n$ transactions. The pipeline solution needs 5 minutes.

4   Experimentation Based on the Copernic Project

Our experimental setup was based on dual processor Xeon systems (each processor is single-core) taking advantage of hyperthreading. Each system is able to run 4 threads in parallel (each processor may run two threads concurrently). Our API has been applied onto a modified version of the Copernic Framework [2]. Operating Systems were based on RedHat Advance Server 4 with JAVA applications running thanks to Sun JDK 1.5.0.9 and Jomp API 1.0b. The first system is dedicated to service, provided by JBOSS 4.0.4-GA, while the other one is on the client side. Because our work focuses on the client side, the server application has been built in such a way it is not a bottleneck. However, it was possible to tune the server response time.

Not surprisingly, parallelizing web service calls lead to major performance improvements on our multithreading-capable systems. For transactions with very small response time, the cost of thread management and probably the limitations

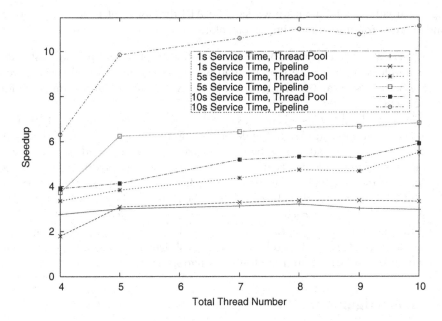

Fig. 7. Speedup for Various Waiting Times and Solutions

of the hyperthreading technology limits the global speedup to a factor of 3.4 instead of an ideal factor of 4. The optimized pipeline solution gives by far the best results compared to the naive scheme.

Our results are depicted in Figure 7. We compared the task pool and the optimized pipeline solutions for various response delays and number of threads. We consider for each experiment 100 service calls with 4000 data to serialize/de-serialize and a response time of 1s, 5s and 10s. The reference speed for speedup computation is the naive scheme using a single thread. For small response times and at least 5 threads, the pipeline solution is a 10% improvement over task pool solution thanks to a lighter thread management cost. The speedup of the pipeline solution over task pool solution grows dramatically with response waiting time, from 30% for 5s response time to more than 60% for 10s response time, since the optimized solution is less challenged by blocked threads.

5   Conclusion and Perspectives

In this paper, we propose an API to efficiently parallelize web service calls. This API is mainly based on a software pipeline which splits web service calls into several tasks, taking advantage of OpenMP for a better use of resources. We have shown that any web service can benefit from our API (both synchronous and asynchronous ones). Using OpenMP to design such an API showed high productivity with slight impact on the original code. To analyze which level of performance and productivity can be reached using different solutions based on, e.g., pthreads is left for future works. The API development is still in progress, and there is room for many improvements:

- Others stages could be added in the pipeline to perform optional tasks: for example, a web service search stage which would aim to find specific web services in several UDDI registries, or an XML data compression stage for reducing the network load.
- The stage load between is not necessarily balanced. For example, if the web service responses are large, the deserialization stage load will be higher than the others. The load can be balanced by setting more threads to this stage. However, the problem is that each web service call can be different, thus the load balancing should be changed at each call. One approach could be the load balancing prediction by estimating the response sizes. Another approach could be to consider each couple (web service task, stage) as a global task in a task pool. In this case, the load does not depend on the pipeline stages but on the threads (cores) used by this task pool. Developing and experimenting with these solutions are the object of ongoing work.

Acknowledgements

The authors would like to thank the IWOMP 2007 anonymous reviewers for their help in improving the quality of the paper.

References

1. Bull, J., Kambites, M.: JOMP, an OpenMP-like interface for Java. In: Proc. of the ACM 2000 Conf. on Java Grande, San Francisco, pp. 44–53 (2000)
2. Direction Générale des Impôts: The copernic tax project (2006), http://en.wikipedia.org/wiki/Copernic_tax_project
3. Tidwell, D.: Web services, the web's next revolution. In: IBM developerWorks (2000)
4. Consortium, W.W.W.: Web services description language (wsdl) (2001)
5. Balart, J., Duran, A., Gonzalez, M., Martorell, X., Ayguade, E., Labarta, J.: Experiences parallelizing a web server with OpenMP. In: First International Workshop on OpenMP (IWOMP 2005), Eugene, Oregon (2005)
6. Kut, A., Birant, D.: An approach for parallel execution of web services. In: ICWS 2004. Proceedings of the IEEE International Conference on Web Services (ICWS 2004), Washington, DC, USA, pp. 812–813. IEEE Computer Society Press, Los Alamitos (2004)
7. Wurz, M., Schuldt, H.: Dynamic parallelization of grid enabled web services. In: Sloot, P.M.A., Hoekstra, A.G., Priol, T., Reinefeld, A., Bubak, M. (eds.) EGC 2005. LNCS, vol. 3470, pp. 173–183. Springer, Heidelberg (2005)
8. Sun Microsystems: Web Services made easier: the Java APIs and architectures for XML. Sun Microsystems (2002), http://java.sun.com/xml/webservices.pdf

Distributed Implementation of OpenMP Based on Checkpointing Aided Parallel Execution

Éric Renault

GET / INT — Samovar UMR INT-CNRS 5157
91011 Évry, France
`eric.renault@int-evry.fr`

Abstract. Checkpointers are used to secure the execution of sequential and parallel programs. This article shows that they can also be used to generate a parallel program from a sequential program automatically, this program being executed on any kind of distributed parallel system. The article also presents how this new technique can be included inside the usual compilation chain to provide a distributed implementation of OpenMP. Finally, some performance measurements are discussed.

1 Introduction

Radical changes in the way of taking up parallel computing has operated during the past years, with the introduction of cluster computing [1], grid computing [2], peer-to-peer computing [3]... However, if platforms have evolved, development tools remain the same. As an example, HPF [4], PVM [5], MPI [6] and more recently OpenMP [7] have been the main tools to specify parallelism in programs (especially when supercomputers were the main issue for parallel computing), and they are still used in programs for cluster and grid architectures. Somehow, this shows that these tools are generic enough to follow the evolution of parallel computing. However, developers are still required to specify almost every information on when and how parallel primitives (for example sending and receiving messages) shall be executed.

Many works [8,9] have been done in order to automatically extract parallel opportunities from sequential programs in order to avoid developers from having to deal with a specific parallel library, but most methods have difficulties to identify these parallel opportunities outside nested loops. Recent research in this field [10,11], based on pattern-maching techniques, allows to substitute part of a sequential program by an equivalent parallel subprogram. However, this promising technique must be associated an as-large-as-possible database of sequential algorithm models and the parallel implementation for any target architectures for each of them.

At the same time, the number of problems that can be solved using parallel machines is getting larger everyday, and applications which require weeks (or months, or even more...) calculation time are more and more common. Thus, checkpointing techniques [12,13,14] have been developed to generate snapshots of applications in order to be able to resume the execution from these snapshots in case of problem instead of restarting the execution from the beginning. Solutions have been developed to resume the execution from a checkpoint on the same machine or on a remote machine,

B. Chapman et al. (Eds.): IWOMP 2007, LNCS 4935, pp. 195–206, 2008.

or to migrate a program in execution from one machine to another, this program being composed of a single process or a set of processes executing in parallel.

This article adresses a different problem. Instead of securing a parallel application using checkpointing techniques, checkpointing techniques are used to introduce parallel computing inside sequential programs, i.e. to allow the parallel execution of parts of a program for which it is known these parts can be executed concurrently. This technique is called CAPE which stands for Checkpointing Aided Parallel Execution. It is important to note that CAPE does not detect if parts of a program can be executed in parallel. We consider it is the job of the developer (or another piece of software) to indicate what can be executed in parallel. CAPE consists in transforming an original sequential program into a parallel program to be executed on a distributed parallel system. As OpenMP already provides a set of compilation directives to specify parallel opportunities, we decided to use the same in order to avoid users from learning yet another API. As a result, our method provides a distributed implementation of OpenMP in a very simple manner.

As presented in this article, CAPE provides three main advantages. First, there is no need to learn yet another parallel programming environment or methodology as the specification of parallel opportunities in sequential programs is performed using OpenMP directives. Second, CAPE inherently introduces safety in the execution of programs as tools for checkpointing are used to run concurrent parts of programs in parallel. Third, more than one node is used only when necessary, i.e. when a part of the program requires only one node to execute (for example if this part is intrinsincly sequential), only one node is used for execution. As performance measurements show, the only drawback of the current implementation is that the checkpointer we used for experiments generates very large images. We are investigating to significantly reduce the overhead involved by the management of these images.

Other works have presented solutions to provide a distributed implementation of OpenMP [15]. Considering that OpenMP has been designed for shared-memory parallel architectures, the first solution was consisting in executing OpenMP programs on top of a distributed shared memory based machine [16]. More recently, other solutions have emerged, all aiming at transforming OpenMP code to other parallel libraries, like Global Arrays [17] or MPI [18].

The article is organized as follows. First, we present CAPE, our method to make a parallel program from a sequential program. Then, we show that the result of the execution of the generated parallel program is equivalent to the execution of the original sequential program. The next section presents how we have developed a distributed implementation of OpenMP on top of CAPE. The last section provides the performance results we have measured on one of our clusters.

2   CAPE

CAPE, which stands for Checkpointing Aided Parallel Execution, consists in modifying a sequential program (for which parts are recognized as being executable in parallel) so that instead of executing each part the one after the other one on a single machine, parts are automatically spread over a set of machines to be executed in parallel. As lots of works are on the way for distributing a set of processes over a range of machines, this article concentrates on how to transform the sequential piece of code to a parallel code. Thus, in the following, we consider that another application (like Globus [19,20], Condor [21,22] or XtremWeb [23,24]) is available to start processes on remote nodes and get their results from there. In this section, this application is called the "dispatcher". CAPE is based on a set of six primitives:

- create (filename) stores in file "filename" the image of the current process. There are two ways to return from this function: the first one is after the creation of file "filename" with the image of the calling process; the second one is after resuming the execution from the image stored in file "filename". Somehow, this function is similar to the fork ( ) system call. The calling process is similar to the parent process with fork ( ) and the process resuming the execution from the image is similar to the child process with fork ( ). Note that it is possible to resume the execution of the image more than once; there is no such equivalence with the fork ( ) system call. The value returned by this function has a similar meaning as those of the fork ( ) system call. In case of error, the function returns -1. In case of success, the returned value is 0 if the current execution is the result of resuming its execution from the image and a strictly positive value in the other case (unlike the fork ( ) system call, this value is not the PID of the process resuming the execution from the image stored in the file).
- diff (first, second, delta) stores in file "delta" the list of modifications to perform on file "first" in order to obtain file "second".
- merge (base, delta) applies on file "base" the list of modifications from file "delta".
- restart (filename) resumes the execution of the process which image was previously stored in "filename". Note that, in case of success, this function never returns.
- copy (source, target) copies the content of file "source" to file "target".
- wait_for (filename) waits for any merges required to update file "filename" to complete.

The following presents two cases where CAPE can be applied automatically with OpenMP: the first case is the "parallel sections" construct and the second case is the "parallel for" construct.

In order to make the piece of code in Fig. 1 readable, operations for both the modified user program and the dispatcher have been included, and operations executed by the dispatcher are *emphasized*. Operations performed by the dispatcher are initiated by the modified user program while sending a message to the dispatcher. Operations are then performed as soons as possible but asynchronously regarding the modified user program.

Error cases (especially when saving the image of the current process) are not represented in Fig. 1(c). In some cases, a test is performed just after the creation of a new image in order to make sure it is not possible to resume the execution from the image direcly. This kind of image is used to evaluate variables updated by a specific piece of code and its execution is not intended to be effectively resumed.

The "Parallel Sections" Construct. Let P_1 and P_2 be two parts of a sequential program that can be executed concurrently. Fig. 1(a) presents the typical code one should write

OpenMP templates

```
# pragma omp parallel sections
{
        # pragma omp section
        P₁
        # pragma omp section
        P₂
}
```

(a) Parallel sections construct

```
# pragma omp parallel for
for ( A ; B ; C )
        D
```

(b) Parallel for construct

CAPE templates

```
parent = create ( original )
if ( parent )
        copy ( original, target )
        ssh hostₓ restart ( original )
        P₁
        parent = create ( after₁ )
        if ( parent )
                diff ( original, after₁, delta₁ )
                merge ( target, delta₁ )
                wait_for ( target )
                restart ( target )
else
        P₂
        parent = create ( after₂ )
        if ( parent )
                diff ( original, after₂, delta₂ )
                merge ( target, delta₂ )
```

(c) Parallel sections construct

```
parent = create ( original )
if ( ! parent )
        exit
copy ( original, target )
for ( A ; B ; C )
        parent = create ( beforeᵢ )
        if ( parent )
                ssh hostₓ restart ( beforeᵢ )
        else
                D
                parent = create ( afterᵢ )
                if ( ! parent )
                        exit
                diff ( beforeᵢ, afterᵢ, deltaᵢ )
                merge ( target, deltaᵢ )
                exit
parent = create ( final )
if ( parent )
        diff ( original, final, delta )
        wait_for ( target )
        merge ( target, delta )
        restart ( target )
```

(d) Parallel for construct

Fig. 1. Templates for both OpenMP and CAPE

when using OpenMP and Fig. 1(c) presents the code to substitute to run P_1 and P_2 in parallel with CAPE.

The first step consists in creating an "original" image used to resume the execution on a distant node, calculate the delta for each part executed in parallel and build the "target" image to resume the sequential execution at the end.

The second step consists in executing parts and generating deltas. Thus, the local node asks the dispatcher to resume the execution of the "original" image on a distant node. Parts are executed, two "after" images are generated to produce two "delta" files; then, these "delta" files are merged to the "target" image; all these operations are executed concurrently. The main difficulty here is to make sure that both the current frame in the execution stack and the set of processor registers are consistent. However, this can be easily achieved using a good checkpointer.

The last step consists in making sure all "delta" files have been included in the "target" image and then restarting the "target" image in the original process.

The "Parallel for" Construct. In the case of for loops (see Fig. 1(b) and Fig. 1(d)), the first step consists in creating an "original" image for both having a reference for the "final" image and having a basis for the "target" image which is used at the end to resume the execution of the program sequentially after having executed the whole loop in parallel. Thus, once created, the "original" image is copied to a "target" image which will be updated step by step using deltas after the execution of each loop iteration.

The second step consists in executing loop iterations. After executing instruction A and checking condition B is true, a "before" image is created. When returning from function create (before), two cases are possible: either the function returns after creating the image, or the function returns after resuming from the image. In the first case, the dispatcher is asked to restart the image on a remote part and then the execution is resumed on the next loop iteration, i.e. executing C, evaluating B... In the second case, the loop iteration is executed and another image is saved in order to determine what are the modifications involved by the execution of this loop iteration (these modifications are then merged asynchronously to the "target" image by the dispatcher).

Once all loop iterations have been executed, the third step consists in creating the "final" image which difference from the "original" image is used in order to set in image "target" modifications involved when executing C and evaluating B for the last time. Moreover, it ensures that the current frame in the execution stack is correct. This is the reason why it is necessary to wait for all other merges to be performed before including the one from "final". When restarting image "target", the execution resumes after create (final). As the process resumes its execution from an image, the last four lines in Fig. 1(d) are not executed the second time.

3   Proof of Concept

In order to prove that our solution is correct, one must show that when executing the program, modified or not by CAPE, the result is the same. That is, the set of updated variables (and the associated values) in the original program is the same for both executions. Two assumptions have to be made.

The first assumption is that the implementation of functions related to image management do not involve onboard effects on the rest of the program. That is, the call to these functions and data structures used by these functions do not change the behaviour of the original program. In fact, functions dedicated to image management can be understood as a transparent set of services provided to the application, and executing a program with or without CAPE must provide the same result.

This assumption is not irrealistic as some checkpointers can be dynamicly linked to the program on which checkpoints have to be performed. In this case, creating an image and resuming the execution from an image can be performed from outside the program, having the program (and its behaviour) unchanged. Thus, both create ( ) and restart ( ) functions can be considered as onboard-effect free. The other four functions (diff ( ), merge ( ), copy ( ) and wait_for ( )) are only dealing with files and no specific data structure is used inside the program. As a result, it is correct to consider that functions related to image management do not involve onboard effects on the user program.

The second assumption is that the parts of the program which are executed in parallel satisfy Bernstein's conditions. In fact, CAPE does not aim at detecting parallel opportunities in sequential programs. It aims at executing parts of a sequential program on a set of nodes of a parallel machine. In order to do so, parts that can be executed in parallel must be identified either by the programmer or by another piece of software. Thus, in the following, we assume that parts to be executed in parallel satisfy Bernstein's conditions.

Let I_i be the set of variables read when executing part P_i and O_i be the set of variables written when executing part P_i. Note that in this context, a "variable" shall be understood in the most general way, i.e. as a "memory location". According to Bernstein's conditions, both P_1 and P_2 can be executed concurrently if and only if all following three conditions are satisfied: $I_1 \cap O_2 = O_1 \cap I_2 = O_1 \cap O_2 = \emptyset$.

The first two Bernstein's conditions indicate that P_1 and P_2 can be started independently, i.e. none requires the other one to be executed before starting its own execution. The third Bernstein's condition, indicates that no updated data is shared by P_1 and P_2. This means that the "target" image (the one used to resume the execution after the parallel execution of P_1 and P_2) can be built in any order by updating data with the content of variables modified during the execution of P_1 and P_2.

Moreover, when an image is created just before the execution of the piece of code P_i and another one just after, the difference between both (the "delta$_i$" computed with function diff ( )) stores the list of all variables updated during the processing of P_i. In other words, file "delta$_i$" represents O_i for part P_i.

Then, as "delta$_i$" files are merged to the original image to produce the "target" image and as, according to the third Bernstein's condition, there is no variable shared by two "delta" files, the "target" image contains all updates performed by each part.

Thus, as no part requires data from the other one (first two Bernstein's conditions), as no variable is updated by both P_1 and P_2 (third Bernstein's condition), and as each modification involved by the execution of both P_1 and P_2 is taken into account before returning in the sequential execution, executing in this context a program sequentially or in parallel with CAPE provides the same result.

Bernstein's conditions are also used to define the concurrency relation ($P_1 \| P_2$ if and only if P_1 and P_2 satisfy Bernstein's conditions). It is then possible to extend the concurrency relation to a larger number of P_i with the following relation:

$$P_1 \| P_2 \| ... \| P_n \quad \equiv \quad P_i \| P_j \quad \forall (i, j) \in [1, n]^2 \quad i \neq j$$

It is also possible to generalize our method to any number of P_i in the same way. This is especially interesting for loops.

4   Distributed Implementation of OpenMP Using CAPE

In order to valid the concepts associated to CAPE, we are developing a distributed implementation of OpenMP. The current implementation is based on top of CKPT version 1.3 [12]. It has been necessary to slightly patch the original version of CKPT so as to be able to make the difference between the execution following the storage of the image of the current process in a file and the execution which is the result of resuming the execution from an image. No other checkpointer has been tried yet. However, we believe it shall be easy to implement this solution on top of any checkpointer as long as functions presented in Sec. 2 are implemented.

DOMPCC (the Distributed OpenMP compiler we developed) is built on top of GCC version 3.2.2. It consists in adding an extra stage in the usual compilation chain for C programs. As shown on Fig. 2, the extra compilation stage (SC, which stands for Specific Compiler) has been added after the C preprocessor (CPP) and before the stage of compilation itself (CC).

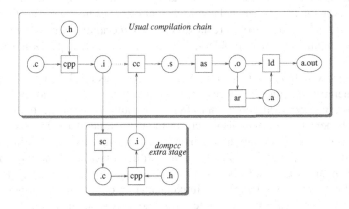

Fig. 2. The compilation chain for DOMPCC

Including the extra compilation stage at this location in the compilation chain allows to take benefits of the result of the C preprocessor (file inclusions, macros, conditionals) and thus to work on a complete C program free of lines beginning with a pound sign (except lines beginning by "# pragma omp" used to identify OpenMP directives). After transforming the original program using SC, the generated ".c" file is processed by the C preprocessor again in order to return in the usual compilation chain at the stage where the usual compilation chain was rerouted.

It is important to note that, as DOMPCC is based on GCC, any option of GCC is naturally available for DOMPCC. For example, if this implementation has to be included in a larger application, it is possible to use DOMPCC instead of GCC for any compilation step. This way, paths to header files and libraries are set correctly and others if any for larger applications can be added conveniently. Once compiled, the executable file is autonomous and can be run directly.

At present, not all the constructs of OpenMP have been implemented and only "parallel sections" and "parallel for" constructs are available. The decision to focus on these two constructs first is based on the fact that they represent the main cases for parallel applications. However, considering there is no technology lock for the implementation of the other constructs, they should be available soon.

5 Discussion

There are many advantages to implement an OpenMP compiler for distributed-memory machine with CAPE and there are also drawbacks. This section presents the main benefits of using CAPE and also the main problems, whereas these problems are inherent to the CAPE (ie. our implementation on top of a distributed memory system) or not.

5.1 Advantages

Use of Checkpoints. The most obvious advantage of using CAPE for an implementation of OpenMP on top of a distributed memory architecture is the use of checkpoints. Even if checkpoints are not used to secure the execution of the application and avoid restarting from the beginning in case of problem, they are still available to perform this task.

Indirect Access to Memory Taken into Account. When using OpenMP on top of a distributed memory machine, it is necessary to exchange data between nodes. With CAPE, exchanges are performed before and after the parallel parts inside the program; for other implementations it may also occur inside the parallel parts. In order to perform the transfer of these information, it is mandatory to determine the location of the corresponding memory area. This may be a very difficult task for some implementations as updated memory areas may be accessed through indirections (ie. using pointers typically). This is not a problem for an implementation on top of a distributed-memory architecture using CAPE as the whole virtual address space is scanned to determine whether a memory area has been updated or not.

Resources used "on demand". Unlike shared-memory systems for which it is very difficult to add resources (especially CPUs) when needed while an application is running, distributed-memory systems are usually quite flexible. With the implementation of OpenMP with CAPE, it is possible to run different parallel parts of a program with various number of nodes. This means that with CAPE it is possible to use only needed resources to run a program.

No Stack Size Limitation. In a shared-memory implementation of OpenMP, all threads are sharing the same virtual address space. This means that all stacks (one for each thread) have to be stored in a single virtual address space. As a result, as one way or another stacks have to be stored the one after the other one, the size of stacks is limited. There is no such a limitation with a implementation of OpenMP on a distributed memory system as whatever the number of threads, there is always a single stack for each virtual address space.

5.2   Drawbacks Inherent to CAPE

Cannot use Shared Variables.   In order to ensure the consistency of the memory distributed on a set of machines, we assumed that all parallel parts of the OpenMP program satisfy Bernstein's conditions. However, this does not allow to take benefits of shared variables. As a result, we need a mechanism to enable all threads to take into account any modifications operated on shared variables.

5.3   Drawbacks Inherent to the Implementation

Excessive use of Bandwidth.   At present, complete checkpoints are sent over the network, first to provide each node all the information needed to restart the process for a specific region of the parallel part and second to return to the master node the result of the execution. In order to reduce the use of the network bandwidth, we have planned to transfer delta files only and no checkpoint file anymore.

Computation of Delta Files.   In the current implementation, the computation of delta files is performed after the generation of checkpoints. This means that two checkpoint files are scanned at the same time to determine the difference between both. To reduce the latency to get the list of modifications that occurred during the execution, we have planned to use an incremental checkpointer instead of the checkpointer we are using that systematically saves the whole virtual address space for each checkpoint.

No Support for Fortran Programs.   At present, only the C programming language can take benefits of our work. A Fortran implementation is scheduled in the future.

6   Performance Evaluation

The performance have been measured on a platform composed of a set of eight Pentium-III running at 800 MHz with 1.2 GB of memory on each node and operated by Linux RedHat version 3.2.2-5 (using Linux kernel 2.4.20-8). The interconnexion network on this platform is either Ethernet 100 Mbit/s or Myrinet 2000. However, our implementation is intended to run on any distributed parallel system, we used only the Ethernet network so as to be as generic as possible.

In order to measure performance, we used a matrix-matrix product of square matrices. The matrix-matrix product has not been optimized. Matrices are dense and each value in the resulting matrix is the sum of the scalar products of the corresponding lines and rows. Table 1 provides the size of the matrices used for the performance evaluation, the total number of elements in each matrix and the size of image files.

Fig. 3(a) presents the speedup for various matrix size. Performance measurements show that the larger the size of the matrix the higher the speedup. In fact, as the complexity of the matrix-matrix product is $O(n^3)$, the larger the matrices, the less important both the network latency to transfer images (which complexity is $O(n^2)$) and the time to determine the set of updated variables (which complexity is also $O(n^2)$). As a result, performance measurements show that, with the current implementation, this technique

Table 1. Matrix and checkpoint size

Matrix size	Number of elements	Checkpoint size
840×840	705 600 → 0.7 Me	10 274 204 → 9.8 MB
1680×1680	2 822 400 → 2.8 Me	35 677 596 → 34.0 MB
2520×2520	6 350 400 → 6.3 Me	78 009 759 → 74.4 MB
3360×3360	11 289 600 → 11.3 Me	137 282 972 → 130.9 MB

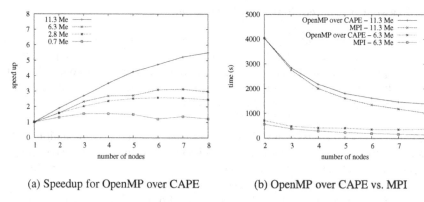

(a) Speedup for OpenMP over CAPE   (b) OpenMP over CAPE vs. MPI

Fig. 3. Performance evaluation

is well-adapted to coarse-grain parallel loops. Moreover, performance show that the larger the size of the grain of parallelism, the better the speed up.

The comparison with an equivalent MPI program is interesting. Fig. 3(b) presents execution times for matrix-matrix products with both CAPE and MPI. The MPI program was written for the experiment and satisfies the same requirements as for CAPE (for the complexity essentially). Performance measurements have been done on the same platform with similar experimental conditions (especially average load for CPUs). MPI is MPICH version 1.2.5.10 with driver CH_P4. Fig. 3(b) shows that, even if the execution time with MPI is always faster than the execution time with CAPE), the time difference between the execution of the MPI program and the program automatically generated by CAPE from the sequential version is not very large.

At present, the main part of the overhead when using CAPE is in the image management. According to Fig. 1(c) and Fig. 1(d), every time an image is generated, it is written on the disk; then, "after" images are compared to the "original" image and the difference is also stored on the disk. A significant improvement could be achieved while using an incremental checkpointer that would directly generates "delta" data instead of "after" images, avoiding the cost to evaluate the difference with the "original" image.

7   Conclusion

This article presented a new way of transforming a sequential program into a parallel program using checkpointing techniques. We showed that CAPE is consistent, i.e. executing a program tranformed using CAPE and executing the original program provide the same result. However, executing a program with CAPE provides three main advantages: first, no specific development is required to develop a distributed-memory parallel version of a program from a sequential version or a shared-memory parallel version; second, as an image is created for each part to be executed concurrently and as no message is exchanged between nodes, it is easier for CAPE to recover in case of problem; third, CAPE is more flexible with the number of nodes required for execution (nodes are used on demand, i.e. only one node is used for the execution of the sequential parts of the program and the number of nodes required for execution may change during the execution).

Then, we presented the distributed implementation of OpenMP we have developed using CAPE. Performance measurements show that it is interesting to execute coarse-grain parallel applications and that the larger the size of the grain, the higher the speed up. Performance measurements also showed that the execution time for large matrices with our implementation is quite similar to the execution time when using MPI. Investigations show that the difference between both execution times is mainly due to the overhead involved by the management of the images and we proposed a solution to investigate in order to bypass the problem.

References

1. Buyya, R.: High Performance Cluster Computing: Architectures and Systems, vol. 1. Prentice-Hall, Englewood Cliffs (1999)
2. Foster, I., Kesselman, C., Tuecke, S.: The Anatomy of the Grid: Enabling Scalable Virtual Organizations. The International Journal of High Performance Computing Applications 15(3), 200–222 (2001)
3. Leuf, B.: Peer to Peer. Collaboration and Sharing over the Internet. Addison-Wesley, Reading (2002)
4. Loveman, D.B.: High Performance Fortran. IEEE Parallel & Distributed Technology: Systems & Applications 1(1), 25–42 (1993)
5. Geist, A., Beguelin, A., Dongarra, J., Jiang, W., Manchek, R., Sunderam, V.S.: PVM: Parallel Virtual Machine: A Users' Guide and Tutorial for Network Parallel Computing. Scientific and Engineering Computation Series. MIT Press, Cambridge (1994)
6. Snir, M., Otto, S., Huss-Lederman, S., Walker, D., Dongarra, J.: MPI: The Complete Reference, 2nd edn. Scientific and Engineering Computation Series. MIT Press, Cambridge (1998)
7. OpenMP Architecture Review Board: OpenMP Application Program Interface, Version 2.5 Public Draft (2004)
8. Allen, J.R., Callahan, D., Kennedy, K.: Automatic Decomposition of Scientific Programs for Parallel Execution. In: Proceedings of the 14th ACM SIGACT-SIGPLAN symposium on Principles of programming languages, Munich, West Germany, pp. 63–76. ACM Press, New York (1987)
9. Feautrier, P.: Automatic parallelization in the polytope model. In: Perrin, G.-R., Darte, A. (eds.) The Data Parallel Programming Model. LNCS, vol. 1132, pp. 79–103. Springer, Heidelberg (1996)
10. Barthou, D., Feautrier, P., Redon, X.: On the Equivalence of Two Systems of Affine Recurrence Equations. In: Monien, B., Feldmann, R.L. (eds.) Euro-Par 2002. LNCS, vol. 2400, pp. 309–313. Springer, Heidelberg (2002)

11. Alias, C., Barthou, D.: On the Recognition of Algorithm Templates. In: Knoop, J., Zimmermann, W. (eds.) Proceedings of the 2nd International Workshop on Compiler Optimization meets Compiler Verification, Warsaw, Poland, pp. 51–65 (2003)
12. Web page: Ckpt (2005), http://www.cs.wisc.edu/~zandy/ckpt/
13. Osman, S., Subhraveti, D., Su, G., Nieh, J.: The Design and Implementation of Zap: A System for Migrating Computing Environments. In: Proceedings of the 5th USENIX Symposium on Operating Systems Design and Implementation, Boston, MA, pp. 361–376 (2002)
14. Plank, J.S.: An Overview of Checkpointing in Uniprocessor and Distributed Systems, Focusing on Implementation and Performance. Technical Report UT-CS-97-372, Department of Computer Science, University of Tennessee (1997)
15. Merlin, J.: Distributed OpenMP: extensions to OpenMP for SMP clusters. In: 2nd European Workshop on OpenMP (EWOMP 2000), Edinburgh, UK (2000)
16. Karlsson, S., Lee, S.W., Brorsson, M., Sartaj, S., Prasanna, V.K., Uday, S.: A fully compliant OpenMP implementation on software distributed shared memory. In: Sahni, S.K., Prasanna, V.K., Shukla, U. (eds.) HiPC 2002. LNCS, vol. 2552, pp. 195–206. Springer, Heidelberg (2002)
17. Huang, L., Chapman, B., Liu, Z.: Towards a more efficient implementation of OpenMP for clusters via translation to global arrays. Parallel Computing 31(10-12), 1114–1139 (2005)
18. Basumallik, A., Eigenmann, R.: Towards automatic translation of OpenMP to MPI. In: Proceedings of the 19th annual international conference on Supercomputing, pp. 189–198. ACM Press, Cambridge (2005)
19. Foster, I., Kesselman, C.: Globus: A Metacomputing Infrastructure Toolkit. The International Journal of Supercomputer Applications and High Performance Computing 11(2), 115–128 (1997)
20. Web page: Globus (2007), http://www.globus.org/
21. Litzkow, M., Livny, M., Mutka, M.: Condor - A Hunter of Idle Workstations. In: The 8th International Conference on Distributed Computing Systems, San Jose, CA, pp. 104–111. IEEE Computer Society Press, Los Alamitos (1988)
22. Web page: Condor (2007), http://www.cs.wisc.edu/condor/
23. Fedak, G., Germain, C., Néri, V., Cappello, F.: XtremWeb: A Generic Global Computing System. In: Buyya, R., Mohay, G., Roe, P. (eds.) Proceedings First IEEE/ACM International Symposium on Cluster Computing and the Grid, Brisbane, Australia, pp. 582–587. IEEE Computer Society Press, Los Alamitos (2001)
24. Web page: XTremWeb (2006), http://www.xtremweb.org/

Author Index

Lecture Notes in Computer Science

Sublibrary 1: Theoretical Computer Science and General Issues

For information about Vols. 1– 4726
please contact your bookseller or Springer

Vol. 4935: B. Chapman, W. Zheng, G.R. Gao, M. Sato, E. Ayguadé, D. Wang (Eds.), A Practical Programming Model for the Multi-Core Era. XII, 207 pages. 2008.

Vol. 4934: U. Brinkschulte, T. Ungerer, C. Hochberger, R.G. Spallek (Eds.), Architecture of Computing Systems – ARCS 2008. XI, 287 pages. 2008.

Vol. 4927: C. Kaklamanis, M. Skutella (Eds.), Approximation and Online Algorithms. X, 289 pages. 2008.

Vol. 4926: N. Monmarché, E.-G. Talbi, P. Collet, M. Schoenauer, E. Lutton (Eds.), Artificial Evolution. XIII, 327 pages. 2008.

Vol. 4921: S.-i. Nakano, M.. S. Rahman (Eds.), WAL-COM: Algorithms and Computation. XII, 241 pages. 2008.

Vol. 4919: A. Gelbukh (Ed.), Computational Linguistics and Intelligent Text Processing. XVIII, 666 pages. 2008.

Vol. 4917: P. Stenström, M. Dubois, M. Katevenis, R. Gupta, T. Ungerer (Eds.), High Performance Embedded Architectures and Compilers. XIII, 400 pages. 2008.

Vol. 4915: A. King (Ed.), Logic-Based Program Synthesis and Transformation. X, 219 pages. 2008.

Vol. 4912: G. Barthe, C. Fournet (Eds.), Trustworthy Global Computing. XI, 401 pages. 2008.

Vol. 4910: V. Geffert, J. Karhumäki, A. Bertoni, B. Preneel, P. Návrat, M. Bieliková (Eds.), SOFSEM 2008: Theory and Practice of Computer Science. XV, 792 pages. 2008.

Vol. 4905: F. Logozzo, D.A. Peled, L.D. Zuck (Eds.), Verification, Model Checking, and Abstract Interpretation. X, 325 pages. 2008.

Vol. 4904: S. Rao, M. Chatterjee, P. Jayanti, C.S.R. Murthy, S.K. Saha (Eds.), Distributed Computing and Networking. XVIII, 588 pages. 2007.

Vol. 4878: E. Tovar, P. Tsigas, H. Fouchal (Eds.), Principles of Distributed Systems. XIII, 457 pages. 2007.

Vol. 4875: S.-H. Hong, T. Nishizeki, W. Quan (Eds.), Graph Drawing. XIII, 402 pages. 2008.

Vol. 4873: S. Aluru, M. Parashar, R. Badrinath, V.K. Prasanna (Eds.), High Performance Computing – HiPC 2007. XXIV, 663 pages. 2007.

Vol. 4863: A. Bonato, F.R.K. Chung (Eds.), Algorithms and Models for the Web-Graph. X, 217 pages. 2007.

Vol. 4860: G. Eleftherakis, P. Kefalas, G. Păun, G. Rozenberg, A. Salomaa (Eds.), Membrane Computing. IX, 453 pages. 2007.

Vol. 4855: V. Arvind, S. Prasad (Eds.), FSTTCS 2007: Foundations of Software Technology and Theoretical Computer Science. XIV, 558 pages. 2007.

Vol. 4854: L. Bougé, M. Forsell, J.L. Träff, A. Streit, W. Ziegler, M. Alexander, S. Childs (Eds.), Euro-Par 2007 Workshops: Parallel Processing. XVII, 236 pages. 2008.

Vol. 4851: S. Boztaş, H.-F.(F.) Lu (Eds.), Applied Algebra, Algebraic Algorithms and Error-Correcting Codes. XII, 368 pages. 2007.

Vol. 4848: M.H. Garzon, H. Yan (Eds.), DNA Computing. XI, 292 pages. 2008.

Vol. 4847: M. Xu, Y. Zhan, J. Cao, Y. Liu (Eds.), Advanced Parallel Processing Technologies. XIX, 767 pages. 2007.

Vol. 4846: I. Cervesato (Ed.), Advances in Computer Science – ASIAN 2007. XI, 313 pages. 2007.

Vol. 4838: T. Masuzawa, S. Tixeuil (Eds.), Stabilization, Safety, and Security of Distributed Systems. XIII, 409 pages. 2007.

Vol. 4835: T. Tokuyama (Ed.), Algorithms and Computation. XVII, 929 pages. 2007.

Vol. 4818: I. Lirkov, S. Margenov, J. Waśniewski (Eds.), Large-Scale Scientific Computing. XIV, 755 pages. 2008.

Vol. 4800: A. Avron, N. Dershowitz, A. Rabinovich (Eds.), Pillars of Computer Science. XXI, 683 pages. 2008.

Vol. 4783: J. Holub, J. Žďárek (Eds.), Implementation and Application of Automata. XIII, 324 pages. 2007.

Vol. 4782: R. Perrott, B.M. Chapman, J. Subhlok, R.F. de Mello, L.T. Yang (Eds.), High Performance Computing and Communications. XIX, 823 pages. 2007.

Vol. 4771: T. Bartz-Beielstein, M.J. Blesa Aguilera, C. Blum, B. Naujoks, A. Roli, G. Rudolph, M. Sampels (Eds.), Hybrid Metaheuristics. X, 202 pages. 2007.

Vol. 4770: V.G. Ganzha, E.W. Mayr, E.V. Vorozhtsov (Eds.), Computer Algebra in Scientific Computing. XIII, 460 pages. 2007.

Vol. 4769: A. Brandstädt, D. Kratsch, H. Müller (Eds.), Graph-Theoretic Concepts in Computer Science. XIII, 341 pages. 2007.

Vol. 4763: J.-F. Raskin, P.S. Thiagarajan (Eds.), Formal Modeling and Analysis of Timed Systems. X, 369 pages. 2007.

Vol. 4759: J. Labarta, K. Joe, T. Sato (Eds.), High-Performance Computing. XV, 524 pages. 2008.

Vol. 4750: M.L. Gavrilova, C.J.K. Tan (Eds.), Transactions on Computational Science I. XI, 181 pages. 2008.

Vol. 4746: A. Bondavalli, F. Brasileiro, S. Rajsbaum (Eds.), Dependable Computing. XV, 239 pages. 2007.

Vol. 4743: P. Thulasiraman, X. He, T.L. Xu, M.K. Denko, R.K. Thulasiram, L.T. Yang (Eds.), Frontiers of High Performance Computing and Networking ISPA 2007 Workshops. XXIX, 536 pages. 2007.

Vol. 4742: I. Stojmenovic, R.K. Thulasiram, L.T. Yang, W. Jia, M. Guo, R.F. de Mello (Eds.), Parallel and Distributed Processing and Applications. XX, 995 pages. 2007.

Vol. 4739: R. Moreno Díaz, F. Pichler, A. Quesada Arencibia (Eds.), Computer Aided Systems Theory – EURO-CAST 2007. XIX, 1233 pages. 2007.

Vol. 4736: S. Winter, M. Duckham, L. Kulik, B. Kuipers (Eds.), Spatial Information Theory. XV, 455 pages. 2007.

Vol. 4732: K. Schneider, J. Brandt (Eds.), Theorem Proving in Higher Order Logics. IX, 401 pages. 2007.

Vol. 4731: A. Pelc (Ed.), Distributed Computing. XVI, 510 pages. 2007.

Vol. 4728: S. Bozapalidis, G. Rahonis (Eds.), Algebraic Informatics. VIII, 291 pages. 2007.